UNDERSTANDING AND MANAGING INTEREST RATE RISKS

SERIES IN MATHEMATICAL FINANCE

Series Editors: Zhaohui Chen *(International Monetary Fund)*
Peter G. Zhang *(Chase Manhattan Bank)*

Published

Vol. 1: Understanding and Managing Interest Rate Risks
(Ren-Raw Chen)

Series in Mathematical Finance

Vol. 1

UNDERSTANDING AND MANAGING INTEREST RATE RISKS

Ren-Raw Chen
Rutgers University

Published by

World Scientific Publishing Co. Pte. Ltd.

P O Box 128, Farrer Road, Singapore 912805

USA office: Suite 1B, 1060 Main Street, River Edge, NJ 07661

UK office: 57 Shelton Street, Covent Garden, London WC2H 9HE

Library of Congress Cataloging-in-Publication Data

Chen, Ren-Raw.

Understanding and managing interest rate risks / Ren-Raw Chen.

p. cm. -- (Series in mathematical finance ; v. 1)

Includes bibliographical references and index.

ISBN 9810227515

1. Interest rate risk -- mathematical models. 2. Financial futures -- Mathematical models. 3. Fixed-income securities -- Mathematical models. I. Title. II. Series.

HG6024.5.C47 1996

332.63'23--dc20 96-9289

CIP

British Library Cataloguing-in-Publication Data

A catalogue record for this book is available from the British Library.

This book is printed on acid-free paper.

Printed in Singapore by Uto-Print

FOREWORD BY THE SERIES EDITORS

The world of finance has been changing drastically in recent decades. Sophisticated financial products have been created to satisfy the various needs of investors. Financial analysis has also entered a new stage, making it possible, for example, for hedge funds to operate in multiple market environments, for proprietary trading to benefit from the slightest arbitrage opportunities, and for economists to make more accurate forecasts.

None of the above achievements would have been possible without the proper use of sophisticated mathematical tools. In fact, such tools are indispensable in today's market activities, be they financial analysis, product development, or the pricing of securities. The fast pace of events in the markets also presents a challenge to regulatory authorities, whose work would be impossible without a knowledge of the technicalities of the new financial products in the market.

Mathematical tools alone, however, cannot make things happen. It is the integration of sound economic and financial theories with the appropriate mathematical tools that forms the cornerstone of today's financial modeling. Such integration is the spirit of this book series.

By presenting the problems specific to each area of market activity, the series will help mathematically competent readers see the interface between quantitative tools and the commonly used economic and financial theories. Economic and financial professionals will be better able to master the technical tools required to fully understand and operate their models. The series will cover all aspects of finance, from the forecasting of economic fundamentals to the pricing of derivative products, from corporate finance to international currency markets, and from empirical tests to theoretical analysis.

We welcome initiatives from prospective authors and feedback from readers. Of course, we thank the Publisher for taking on this series. In particular, we wish to thank Dr. K. K. Phua, the Chairman of World Scientific, for his encouragement, and Mr. Yew Kee Chiang and Dr. Richard Lim for their editorial help.

Zhaohui Chen
Washington, D.C.

Peter G. Zhang
Tokyo

PREFACE

Since the macro policy change in October, 1979, interest rates have become floating and how to manage interest rate risks has become a major task for both academics and practitioners. Fixed income research has therefore emerged as a mainstream research topic in finance. Fixed income research differs from those yield curve studies in early 60's in that it emphasizes an equilibrium solution to the yield curve. The ability to identify a reasonable solution enables traders to manage and hedge interest rate risks.

The early work of "modern" term structure theories can be traced back to 1977 when Vasicek and Cox, Ingersoll, and Ross (CIR) developed simultaneously the same model; only that Vasicek adopted a normality assumption while Cox, Ingersoll, and Ross used a scaled non-central chi-squared distribution. This early work has opened a big door for later researchers and resulted in fruitful research in the past 20 years.

Recognizing single factor models being overly simplistic, multi-factor models were later developed to solve the fitting problem of the models by Vasicek and CIR. There are Brennan and Schwartz's model in 1978, Richard's model in 1978, and Langetiegls model in 1980 for representing the early work and Longstaff and Schwartz's model and Chen and Scott's model, both in 1992, for recent developments.

Along with the developments of multi-factor models have been the estimation techniques for the parameters in these models. The regression method, generalized method of moments, maximum likelihood estimation, and most importantly the state space model have been used in estimating multi-factor Vasicek or CIR models.

Taking a slightly different view from the multi-factor models, researchers developed another series of models that take observed yield curve as given; so that fitting becomes never a problem. This goal is accomplished by making the parameters in the Vasicek or Cox–Ingersoll–Ross model time dependent. The early work is the Ho and Lee model in 1986. Dybvig, although never published his work, gives an extension to the Ho–Lee model which is in spirit similar to a published work by Hull and White in 1990. Heath, Jarrow, and Morton in 1992 provided a framework which relates forward rates and spot rates.

Students have long been taught that there is no free lunch in finance. The choice or trade-off between the multi-factor and time-dependent models is a trade-off between simplicity and accuracy. Most time-dependent models do not have easy solutions and the parameters cannot be estimated easily. Multi-factor models cannot provide prefect fits to the yield curve and in turn cannot price interest rate derivatives correctly. In the last chapter, we try to propose an integrated approach so that both goals can be accomplished but its validity needs to be tested.

Since this book emphasizes the theory, it is written for readers who are interested in advanced trading and hedging interest rate derivatives. This includes practitioners such as traders, risk managers, and auditors. Students who are interested to participate in this financial market should also find this book helpful. They are second year MBA and MS students majoring in derivatives. Anyone who has been trained in other professional areas and seeks to enjoy the beauty of financial engineering and derivatives pricing can use this book as an introduction because some materials of this book are written in a quite technical way.

To introduce a whole literature in a logical way, I organize the book as follows: In the first chapter, the basic arbitrage concept is introduced. Some issues about bonds are also discussed in this chapter. In Chapter 2, classical term structure models of Vasicek and Cox–Ingersoll–Ross are introduced. In this chapter, an easy derivation of each model is briefly sketched. Multi-factor models are given a quick look in this chapter as well. Option and futures pricing formulas are presented separately in Chapter 3 because they can span the whole space of interest rate derivative contracts. In Chapter 4, I list a series of common interest rate contracts and their pricing formulas. The purpose of this chapter is twofold. First, it gives an overview of currently traded interest rate contracts. Second, this chapter shows how to apply the pricing methodology in Chapters 2 and 3 to these contracts. Hopefully interested readers can price any other contract they encounter using the techniques introduced in this chapter. Chapter 5 gives details of how the estimation of the fixed parameter models can be done, from the easiest regression to the most difficult state space model with Kalman filtering. In Chapter 6, we demonstrate how hedges should be constructed in both multi-factor models and time-dependent models. Explicit examples are given. Finally, in Chapter 7, an approach that integrates both multi-factor and time-dependent models is proposed. This method is believed to be superior but needs be tested.

The fixed income research has been a fascinating research area and is believed to remain popular for many years to come. New products are being developed every day. I hope that presenting you this book can broaden the audience of this market so that it can continuously grow and efficiently trade.

Ren-Raw Chen

New Brunswick, New Jersey

CONTENTS

SYMBOLS AND NOTATION

α	Speed parameter of the Vasicek model.
κ	Speed parameter of the CIR model.
μ	Reverting level of the Vasicek model.
θ	Reverting level of the CIR model.
σ	Volatility parameter of the Vasicek model, also instantaneous standard deviation of the instantaneous short rate.
υ	Volatility parameter of the CIR model.
q	Market price of interest rate risk in the Vasicek model.
λ	Market price of interest rate risk in the CIR model.
r	Instantaneous short rate.
y_i	State factor i in a multi-factor model.
K	Strike price of an option.
$P(t,T)$	Pure discount bond price at current time t to maturity time T.
$y(t,T)$	Yield to maturity of $P(t,T)$.
$Q(t,\underline{s})$	Coupon bond price at current time t with a vector of coupon arrival times, $\underline{s}$.
$\Psi(t,T_f,T)$	Forward price at current time t of a pure discount bond $P(T_f,T)$
$\Phi(t,T_f,T)$	Futures price at current time t of a pure discount bond $P(T_f,T)$.
$f(t,T_f,T)$	Forward rate of a forward price $\Psi(t,T_f,T)$.
$f(t,T)$	Instantaneous forward rate at time t of an instantaneous forward price $\Psi(t,T,T+dT)$.
$O_c(t,T_c,T)$	Call option written on a pure discount bond $P(t,T)$.
$O_p(t,T_c,T)$	Put option written on a pure discount bond $P(t,T)$
$V(t,T)$	Variance of the bond return, i.e., $\text{var}_t[\ln P(T,s)]$
$R(t,T)$	Total return from t to T, $\int_t^T r(u)du$, or $1+r\Delta t$ when $T=t+\Delta t$.
$u(T-t)$	Perturbation function (up) in the Ho–Lee model.
$d(T-t)$	Perturbation function (down) in the Ho–Lee model.

ACKNOWLEDGMENTS

This book is devoted to my parents, Jer-Wei She Chen and Jaw-Lin Chen.

It is a theoretical introduction to term structure theories as well as a summary of my research for these many years on the term structure modeling and contingent claim pricing. First I would like to thank Dr. Phelilm P. Boyle for arousing my interest in this fascinating area when I was a student at the University of Illinois and for being my mentor for so many years, even after my graduation. I also thank Dr. Louis O. Scott for his patient guidance of my dissertation as well as co-authorship for so many papers that we wrote together. I owe them a great deal which can never be repaid. I want to thank them for making the book possible. Dr. Tyler T. L. Yang, a longtime friend and co-author of my mortgage papers, has helped a great deal in making the book possible. My colleague at the National Taiwan University (Computer Science Department), Dr. Yuh-Dauh Lyuu, who carefully read some material and corrected many mistakes in the initial version, deserves my most sincere gratitude. My doctoral student at NTU, Mr. Yeh Shi-Guo, and MBA students at Katz School of Business of the University of Pittsburgh also corrected numerous mistakes in the original manuscript. The generous editorial help provided by World Scientific is greatly acknowledged. The editor, Mr. Yew Kee Chiang, has been particularly helpful in finalizing the manuscript. Finally, I would like to thank one of the series editors, Dr. Peter Zhang, who encouraged me to write the book. His friendship and support are invaluable.

My apologies go to my wonderful wife, Miranda (Hsing-Yao) Chen, who has already suffered for so many years from her workaholic husband, during the time when the book was written, and to my two children, Janice and Jonathan, whom I should have spent more time with.

1/27/01

CHAPTER 1
BOND PRIMER

1.1 BOND VALUATION AND YIELD TO MATURITY

In most fundamental finance texts, bonds are first valued with the "constant interest rate" or the flat yield curve and then the non-flat or deterministic yield curve. Most texts ignore the most realistic case where interest rates are changing over time for it is too complex for beginners. However, later on, it is inevitable to discuss interest rate fluctuation as an important financial risk. The gap between the theory and the pratice has never been filled until recently.

Thanks to the modern development of interest rate theories in recent years, bond valuation is no longer a difficult issue under changing interest rates environment. The theory of no arbitrage first put forth by Black and Scholes and Ross in 1973 has brought a new thought to bond valuation and interest rate securities. The "no arbitrage" argument now being used makes bond valuation easy and straightforward. Unfortunately, most textbooks, especially those of fundamental finance that are most needed, are still behind in catching modern pricing technology.

What I intend to provide in this book is to introduce the most modern interest rate theories to date in the most straightforward and easiest manner while complex mathematics are kept in appendices. A lot of what I introduce in this book has been well adopted by big Wall Street firms. In fact, a lot of today's pricing concept was initiated by these firms, such as fitting the yield and volatility curves.

1.1.1 Concept of Arbitrage

Since our real world says interest rates are constantly changing, we shall price bonds directly in the world of stochastic interest rates. Before we price bonds, let us first look at an example.

[EXAMPLE 1]
Suppose a one-year T-Bill now traded at 92.31 or $0.9231 for $1 face value.[1] If there is a project that will pay off $500 in certainty in a year,

[1] T Bills are quoted by percents which is a discount yield. To get a traded price from the yield quote, we need to use the following conversion formula:

$$P = 100 - rT$$

where r is the percentage quote and T is time to maturity of the bond in years.

then what is the present value of this project? The answer is:

$500×0.9231=$461.55

This way of finding the present value of a project is different from the usual way that we discount $500 at the riskless rate. Because no rate can be unchanged for a year, the conventional text book approach is not implementable. What we provide above is very implementable because T Bills are traded contracts and very easy because it does not even need discounting in which a problematic discount rate is assumed. We shall show why the present value of a $500 project can not be anything else.

If the present value is less than $461.55, say $450. Then we can short sell 500 units of $1 T Bill to get $461.55. Use only $450 for the project and keep $11.55. At the end of the year, the project will pay $500 in certainty which is used to return the obligation of 500 T Bills. Since this $11.55 is for us to keep at no risk, other investors like us will certainly join and the demand of the project will pull up the price. Sooner or later, the present value will reach an equilibrium value of $461.55. On the other hand, if the present value is more than $461.55, say $470, then we can perform the opposite. Since the market "prices" the project at $470, we, instead of being an investor, can be a supplier of such a project. Anyone who is willing to invest can sign a contract with us; he offers $470 in cash now and we offer $500 in return in a year. By doing so, we get $470 and use only $461.55 to buy 500 T Bills which guarantee $500 payoff in a year. Therefore, we can have $8.45 as an arbitrage profit. As discussed earlier, the free market assumption implies that everyone can be a supplier and the oversupply of such projects will bring down the price and sooner or later, the present value of the project will become $461.55. It is only this value that the supplier and the investor can agree on.

This example, although simple, tells a very important concept — priced by arbitrage. Since we assume everyone in the marketplace is equally smart, there should exists no arbitrage opportunities. A price that permits no arbitrage is a "good" price that everyone in the marketplace can agree on. It has been seen that if an asset is priced by arbitrage, the pricing is simple and the price is unique.

If a project pays out multiple cash flows, the present value is the sum of "discounted" (by T Bills) payoffs. In other words, we multiply each cash flow amount by a corresponding T Bill price and then sum up. Therefore, T Bills are just like discount factors. Finding the present value this way, we get a value that is unique and free of arbitrage. We no longer need to worry if the yield curve is non-flat or if interest rates are stochastic.

[EXAMPLE 2]

Suppose a project pays $500, $400, and $300 in certainty in the next three years. And at the same time we observe three T Strip prices as 92.31, 85.21, and 78.66 for the one, two, and three years to maturity (face value of $1). What is the present value?

As you might have guessed, the answer is:

$$\$500 \times 0.9231 + \$400 \times 0.8521 + \$300 \times 0.7866 = \$1038.37$$

The arbitrage goes as follows. If the present value is not $1038.37, say higher, then we can short sell 500 units of one year T Bills, 400 units of two year T Strips, and 300 units of three year T Strips. The proceeds of the Treasury instruments can be used to invest in the project which will finish the obligation of $500 the first year, $400 the second year, and $300 the third year. The difference is an arbitrage profit. Similarly, the present value should be less than $1038.37 because if so reversing the process can also generate arbitrage profits. Once again, since T bills and T Strips are traded securities, the present value of $1038.37 is unique and simple.

Sometimes, we may find markets be "incomplete", i.e., there are securities that we need for performing arbitrage are not traded in the marketplace. In such a case we need to use synthetic strategies, or close substitutes, or use model prices as proxies. For example, before 1987, long-term zero coupon bonds were not traded. There were no T Strips for us to find the present value. However, coupon bonds were traded and they could be used to back out long-term discount bond prices. these are synthetic strategies. When there are complex contracts to price, synthetic strategies can sometimes be quite complex, some of which may need constant rebalancing of the portfolios, some of which might even need the help of certain models.

In the previous two examples, projects are riskless in a sense that they have no uncertainty risks, i.e., they have no default risks. However, we all know that project do have risks, especially default risks. Therefore, it is inappropriate to use Treasuries to perform arbitrage. Corporate bonds with similar risks, on the other hand, could be better choices and should be used as close substitutes, given that exact instruments are not available.

If none of the above is possible, arbitrage is done with the help of certain models. Models have assumptions and generally are oversimplified. So model prices are not good choices for performing arbitrage strategies. Nonetheless, they still provide a guideline of how contracts should be priced.

Pricing by arbitrage needs traded securities.

1.1.2 Bond Valuation

Given the above background, we know that pricing by arbitrage needs traded securities. More clearly, pricing a contract by arbitrage is to use prices of other traded securities. In our examples, zero coupon Treasuries are used to price investment projects. In this section, we shall continue to use zeros to price other bonds.

While previously mentioned investment projects may not be good examples, Treasury bonds which have 100% certainty payoffs at their maturities are perfect examples. Therefore, the pricing of Treasury coupon bonds is exactly the same as our second example.

[EXAMPLE 3]
From Example 2, we have Treasury zeros priced at 92.31, 85.21, and 78.66 for the one, two, and three years to maturity (face value of $1). Then what is the price of an 8%, three year Treasury coupon bond? The answer is:

$$\$8\times0.9231+\$8\times0.8521+\$108\times0.7866=\$99.15$$

It is clear that this method can be applied to price coupon bonds with all coupons and all maturities.

If coupon bonds are priced with zero coupon bonds, then how are zero coupon bonds priced? Again, if they are priced by arbitrage, we need traded securities. Coupon bonds are the traded securities to price zero coupon bonds. Let us see the following example.

[EXAMPLE 4]
Suppose there are three coupon bonds with one year, two years, and three years to maturity priced at $99.07, $100, and $102.62 (face value of $100) respectively. Their coupon rates are 6%, 7%, and 8% and for simplicity coupons are paid annually. What are the prices for zero coupon bonds for one, two, and three years?

Although the arbitrage process is rather tedious and we shall explain later, the solution is quite simple. We assume P_1, P_2, and P_3 be the three prices. We know that:

$$\begin{cases} 99.07 = 106P_1 \\ 100 = 7P_1 + 107P_2 \\ 102.62 = 8P_1 + 8P_2 + 108P_3 \end{cases}$$

Solving this set of simultaneous equations for three unknowns, we get all three discount bond prices:

$$\begin{cases} P_1 = 0.9346 \\ P_2 = 0.8735 \\ P_3 = 0.8162 \end{cases}$$

The arbitrage process goes as follows. The first coupon bond (6%, one year) is regarded as a pure discount bond but with a slightly different face value, $106. Therefore, its value is simply to scale the face value to $1, which is $0.9346. With this discount bond, we know that we could price the 7%, two year coupon bond if we knew the two year discount bond price. Since we know the coupon bond price already, we can solve for the two-year discount bond price. The following chart explains the relationship.

7 107

7·0.9346

107·?

100

Solve for the question mark, we get 0.8735. The first two discount bond prices are then used, with the third coupon bond price to find the third discount bond price. The logic is explained below:

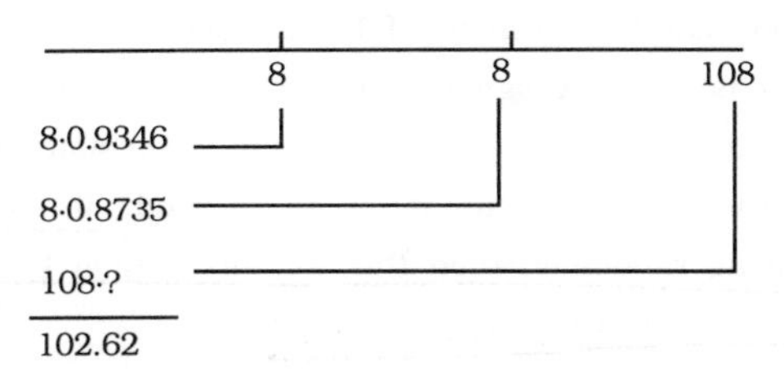

Now, we have seen that discount bonds are used to price coupon bonds and coupon bonds can also be used to price discount bonds. "Priced by arbitrage" actually just depicts no-arbitrage relationships among securities. If certain securities can be linked by no

There is no argument of a bond's price but the yield of a bond is a formula-related matter. A yield tells on average the RoR on a bond.

arbitrage relationships, they can price one another by arbitrage. On the other hand, if they have no arbitrage relationships, then they cannot be priced by arbitrage. Fortunately, most of the financial securities we are interested in have no arbitrage relationships. In this book, we introduce interest rate contracts. Arbitrage pricing is our only pricing methodology.

1.1.3 Yield to Maturity

At most times, the yield, but not the price, of a bond is more commonly used for communication, although price that represents the amount of money necessary to acquire the contract better relates to reality. Therefore, in this section, we present a common concept to calculate "the" yield of a bond. It should be noted that there is no argument of a bond's price since it presents the amount of money required for acquiring the bond but there is great deal of argument of the yield of a bond since the yield is a formula-related number. A yield merely tells on average the rate of return of a bond. It may vary along with the investor's cost of capital. We shall discuss this issue in this section.

The most common formula is a discrete time formula in which the cost of capital assumption is made consistent with the coupon payment times:

$$P = \frac{c/m}{(1+y/m)^1} + \frac{c/m}{(1+y/m)^2} + \cdots + \frac{100+c/m}{(1+y/m)^{m \cdot n}}$$

where P is the coupon bond price that pays c as coupons, m is the frequency of the coupons, and n is the number of years for the bond to mature. In general, m is 2 (semi-annual coupon payments). Throughout this book, the face value for coupon bonds is always assumed as \$100. This is also consistent with the market quotes of the coupon bonds where they represent the percentage of the actual contract size (face value).

This formula is often recognized as the "bond pricing formula" in most introductory finance text books. If the interest rate is constant, then this "yield to maturity" formula is indeed the pricing formula of a bond. If interest rates are not constant but changing randomly, then this formula cannot be used for pricing. However, it is still a valid yield-to-maturity formula.

This formula assumes that the buyer of the bond has a cost of capital that is consistent with the coupon payment interval. The buyer may have a different cost of capital assumption. Then he should use a formula that is consistent with his assumption. For example, if the buyer has to borrow money to buy a bond, then for him to break even, the interests charged by the lender should be matched by the interests earned from the bond. If the lender charges the bond buyer 10% per year fixed for the next 6 years, the bond buyer should be looking for a bond that has the yield to maturity of 10% according to the following formula:

$$P = \frac{c}{(1+y)^1} + \frac{c}{(1+y)^2} + \cdots + \frac{100+c}{(1+y)^n}.$$

The bond price and the coupon should satisfy this formula (with y=10%), not the first one. Of course, one can plug price (P) and coupon (c) into the first equation and find a yield to maturity consistent with the semi-annual compounding. But that is mistaken. In the term structure research, we often assume continuous time compounding because it is more convenient to link to the models developed for interest rate derivatives. The formula is:

$$P = e^{-yt_1}\left(\frac{c}{m}\right) + e^{-yt_2}\left(\frac{c}{m}\right) + \cdots + e^{-yt_n}\left(100 + \frac{c}{m}\right)$$

where the times in the exponential are the time intervals between today and coupon payment dates.

Finally, there is a popular approximation formula for the yield to maturity. It is seen that all formulas have the target variable y on the right hand side, which means there is no easy way to solve for y. When P is plugged in the formula, we need to solve this high order polynomial equation numerically, i.e., through trial and error. So this approximation formula comes in handy:

$$y = \frac{c + \frac{100-P}{n}}{\frac{100+P}{2}}.$$

It can be seen that this formula is *insensitive* to the compounding frequency. This formula has also been proven to have large errors if the bond is not priced close to par.

For the yield to maturity of a pure discount bond, there is no need to iterate for an answer. For the annually discrete yield to maturity of a pure discount bond, we use:

$$y = \sqrt[n]{\frac{1}{P}} - 1$$

where P is the current price of the pure discount bond with a face value of \$1. For other discounting frequency, the formula should be adjusted accordingly. For example, the yield for a bond that expires in 2 years and 1 month needs to be clear about the discounting assumption. If it is discounted monthly, then the above formula should be:

$$y = \left(\sqrt[25]{\frac{1}{P}} - 1 \right) \cdot 12.$$

If it is discounted annually, then there should be an assumption for the discounting of the 1 month. So the discounting is done in two steps. This two-step discounting is common in the industry and there are a number of ways to perform the discounting of the second step. In continuous time, there is no such trouble, we simply use:

$$y(t,T) = -\frac{\ln P(t,T)}{T-t}$$

where $P(t,T)$ is the pure discount bond price at time t that pays \$1 at the maturity time T.

1.2 THE TERM STRUCTURE OF INTEREST RATES OR YIELD CURVE

1.2.1 Shape of the Yield Curve

After we compute yields of different-maturity bonds, we can plot them on a diagram to examine the relationship between the time to maturity and the yield to maturity of various bonds. This relationship is called the term structure of interest rates or more practically, the yield curve. Take the discount bonds from Example 4, we can graph a yield curve for these three bonds.

[EXAMPLE 5]

The three discount bond prices are 0.9346 for one year to maturity, 0.8735 for two years to maturity, and 0.8162 for three years to maturity. To draw their yield, we use a simple assumption of annual compounding. Then the three yields are:

$$\begin{cases} 0.9346 \cdot (1+y_1)^1 = 1 \Rightarrow y_1 = 7\% \\ 0.8735 \cdot (1+y_2)^2 = 1 \Rightarrow y_2 = 7\% \\ 0.8162 \cdot (1+y_3)^3 = 1 \Rightarrow y_3 = 7\% \end{cases}$$

Therefore, the yield curve is flat across different times to maturity. To draw the picture, we conventionally put the time to maturity on the horizontal direction:

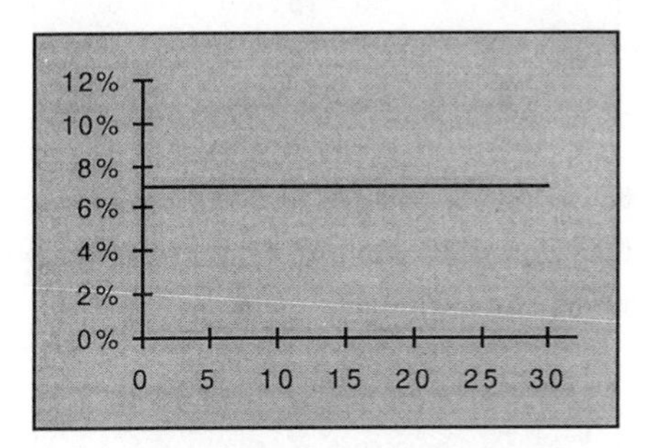

Figure 1. Flat Yield Curve.

As argued earlier, it is unlikely that a real yield curve is flat. Yield curves in general takes a upward sloping shape as follows:

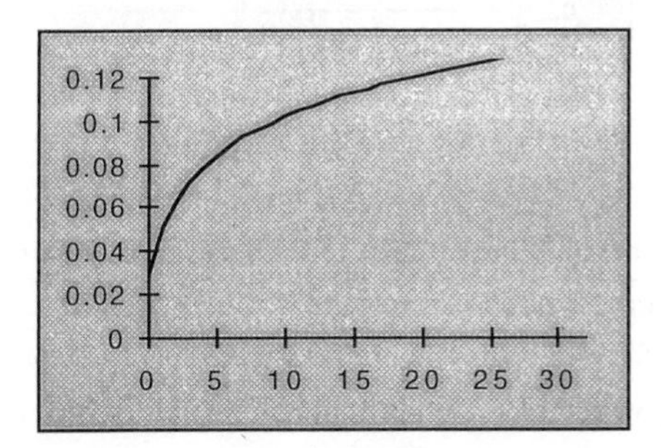

Figure 2. Upward Sloping Yield Curve.

It has been occasionally that the yield curve becomes downward sloping. The graph can look like:

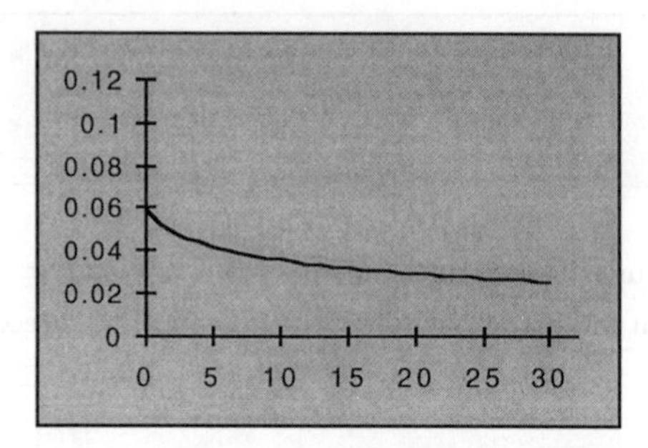

Figure 3. Downward Sloping Yield Curve.

Other than the above two monotonic shapes, sometimes a yield curve can be mixed, i.e., upward sloping and then downward sloping or downward sloping and then upward sloping. This type of curves is called humped shaped:

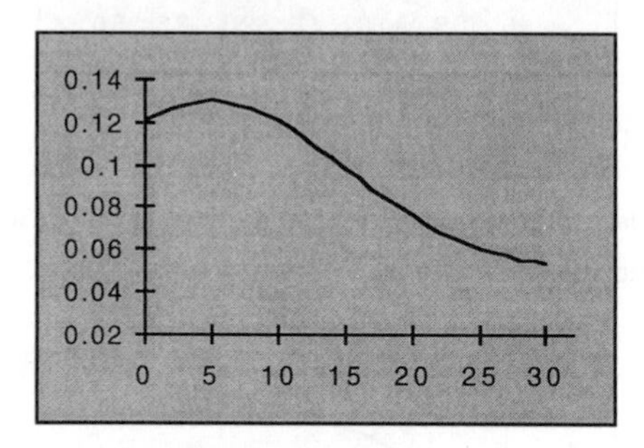

Figure 4. Humped Yield Curve.

The shape of the yield curve used to be believed as a strong indication of how future interest rates might move, like upward sloping yield curve means rising interest rates in the future or forward rates can predict future spot rates. Although modern term structure theories have found that this is untrue, there is still a relationship among future interest rates, risk premia, and the shape of the yield curve. We shall leave this complex issue to later chapters

1.2.2 The Coupon Issue

In the previous section, we use a series of discount bonds to find the yield curve. From the yield curve formula, it is clear that the yields for various bonds calculated by the formula are functions of the coupon. If we fix everything else and change only the coupon, we can get a whole different series of yields. In other words, the yield curve we have must be of the same coupon; or else we are mixing oranges with apples. While the theory introduced in later chapters gives a clearer picture of the underlying reasons, a quick intuition can explain why it is so. Pure discount bonds are believed to have the greatest interest rate risks because they pay no cash until maturity. If you compare a 5-year coupon issue with a 5-year pure discount issue, you will find that the coupon issue has a higher price. The coupon issue can be sold at higher or lower the face value but the discount issue is always sold at the discount price. The higher price paid to the coupon issue is compensated by coupons being received prior to maturity. This high-price–fast-coupon design of the coupon issue should of course reflect lower risk. Therefore, if we draw the yield curve of a coupon bond series, it should lie below that of the discount bond series. After all, there is no free lunch, higher risk should be compensated by higher return.

1.2.3 Forward Rates and Forward Prices

Forward rates are defined as the implied yield to maturity in the future by two current yields to maturity. The following picture explains clearly the forward rates.

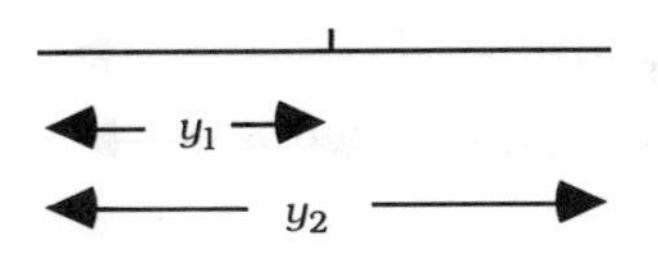

The forward rate of the second year in the above diagram is:

$$(1 + y_1)(1 + f) = (1 + y_2)^2$$
$$f = \frac{(1 + y_2)^2}{1 + y_1} - 1$$

This tells that if you earn y_1 the first year and f the second year, it is equivalent to earning y_2 for 2 years. This rough definition satisfies most needs. In the early days, due to this definition, people regarded forward rates as predictors of future spot rates (yields). Modern theories show that forward rates are really just yields on forward prices. They do not in theory give estimates of future spot rates.

Again we use discount bonds as an example. Suppose we have two bonds, one year and two years to maturity, priced at 0.9434 and 0.8735 respectively. Suppose also that the forward market for discount bonds is open and trading actively. We can consider the following two strategies. We can buy a two-year bond, paying \$0.8735 now and receiving \$1 in two years or we can buy a one-year bond and at the same time lock in a one-year forward contract that delivers a one-year bond in one year. The forward price should be so set that neither strategy can dominate the other in order to avoid arbitrage (priced by arbitrage!)

It is clear that since the forward contract delivers a one-year bond in a year, it will automatically pays \$1 at the end of two years, consistent with the first strategy. If we can spend no more nor less than the cost of the first strategy, \$0.8735, then the two strategies are identical; or arbitrage will take place. Therefore it is ideal for the second strategy to buy a one-year bond in partial units so that it costs only \$0.8735 and then use the proceed to exercise the forward contract. To spend only \$0.8735 for the one-year bond, we can only buy 0.9252 units so that \$0.9434×0.9252 is equal to \$0.8735. The payoff of this transaction is of course \$0.9252 since the face value is one. If the forward price can be set to \$0.9252, then the proceed can exactly be used to get another one-year bond which guarantees \$1 at the end. Therefore, the forward price should be exactly \$0.9252. The following graph gives a clear picture of the process:

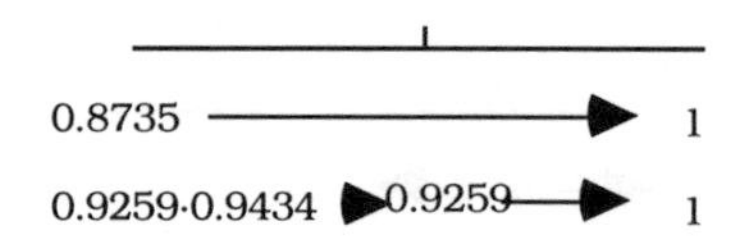

The forward rate is defined as the rate of return of investing \$0.9259 for one year. So, the forward rate is:

$$\$0.9259 \times (1+f) = \$1$$
$$f = 8.03\%$$

This arbitrage concept for forward prices and rates can be extended to any arbitrary periods. To arrive at the general formula, we shall define the bond prices, yields, and forward rates more carefully. We define the following:

$P(t,T) \equiv$ discount bond price at time t that matures at T
$y(t,T) \equiv$ yield to maturity from t to T
$\Psi(t,T_f,T) \equiv$ forward price at time t that delivers at time T_f a T-maturity bond
$f(t,T_f,T) \equiv$ time t forward rate which is the return of $\Psi(t,T_f,T)$.

Then the forward price formula is:

$$\Psi(t, T_f, T) = \frac{P(t,T)}{P(t,T_f)}.$$

The forward rate formula varies. In a simple discrete time setting, we have:

$$f(t, T_f, T) = \sqrt[T-T_f]{\frac{1}{\Psi(t,T_f,T)}} - 1.$$

In continuous time, the forward rate is defined as the instantaneous rate of return which means:

$$T - T_f = \Delta t$$

where Δt is a very small time interval. Then the forward rate notation and its formula are simplified to:

$$f(t, T_f) = -\frac{d \ln P(t, T_f)}{dT_f}$$

which becomes the differentiation of the log of today's bond price with respect to the maturity time. Let us sketch a brief derivation here about this result. We use the complete definition for the forward rate:

$$\begin{aligned} f(t, T_f, T) &= \sqrt[T-T_f]{\frac{1}{\Psi(t,T_f,T)}} - 1 \\ &= \sqrt[T-T_f]{\frac{P(t,T_f)}{P(t,T)}} - 1. \end{aligned}$$

Change this expression to continuous discounting to get:

$$f(t, T_f, T) = -\frac{\ln\left[\frac{P(t,T)}{P(t,T_f)}\right]}{T - T_f} = -\frac{\ln P(t,T) - \ln P(t,T_f)}{T - T_f}.$$

If $T - T_f = \Delta t$, then we reach the definition of the forward rate above.

1.2.4 Price–Yield Relationship

Finally, in this chapter, we present a simple but important result which is the price and yield relationship. From the yield to maturity formula, we learn that the relationship between price and yield is a complicated high-order polynomial function. Nonetheless, in the relevant yield range, say from 0% to 30%, this high-order polynomial function is merely a monotonic convex function. The easiest way to see this is to draw a graph on an example. Let us calculate prices of a 10% coupon, 5 years to maturity bond with various yields. With the yield to maturity formula given above, we obtain the following table.

yield	price	yield	price	yield	price	yield	price
0%	$150.00	8%	$107.99	16%	$80.35	24%	$61.56
1%	$143.68	9%	$103.89	17%	$77.60	25%	$59.66
2%	$137.71	10%	$100.00	18%	$74.98	26%	$57.84
3%	$132.06	11%	$96.30	19%	$72.48	27%	$56.09
4%	$126.71	12%	$92.79	20%	$70.09	28%	$54.42
5%	$121.65	13%	$89.45	21%	$67.81	29%	$52.82
6%	$116.85	14%	$86.27	22%	$65.64	30%	$51.29
7%	$112.30	15%	$83.24	23%	$63.55	31%	$49.82

The graph is like:

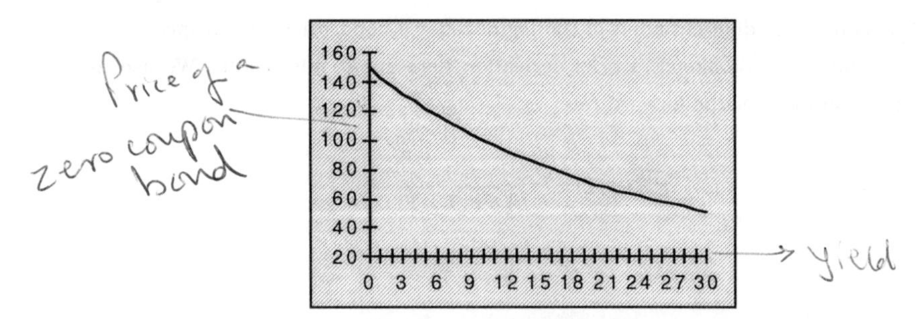

Figure 5. Price–Yield Relationship.

It is noted in the table that when the coupon rate equals the yield, the bond is sold at par. When the yield is higher than the coupon rate, the price is lower than the face value and vice versa. This is called discount or premium sales. We herein reminds our readers a well known result in which for discount issues more frequent payments make

Time to maturity does not correctly reflect the price uncertainty of a bond due to time. Coupon payments reduce the risk so high coupon bonds carry less risk than low coupon bonds

less discount and for premium issues more frequent payments make more premium. The reason is that for premium issues, coupon rate is higher than the yield, which means investors get more from the bond than the market, after the adjustment of time value. Therefore, more frequent coupon payments mean investors get penalized less by time value and the market will ask more for the bond and the bond will be sold at a even higher price. On the other hand, if the bond is sold at discount, since the time value is penalizing the low coupons, more frequent payments suffer even more from the penalization by time value.

The non-linear relationship between price and yield creates a lot of trouble in bond investment and risk management. The concepts of duration and convexity are therefore derived for passive interest rate risk management.

1.5 DURATION AND CONVEXITY

1.5.1 Definition

The concept of duration came up because the time to maturity of a bond does not correctly reflect the price uncertainty due to time. Since coupon payments reduce the risk, high coupon bonds should carry less risk than low coupon bonds. Two bonds, therefore, with same time to maturity will not have the same interest rate risk which is a function of time. Duration is understood as a better measure of the interest rate risk caused by time. Duration has long been used for bond investments and interest risk management. However, duration only provides static management tactics. In a dynamic world, duration is not sufficient to provide reasonable hedges. For a historical reason, we still introduce this interesting concept.

A duration is defined by the following formula which was first proposed by Macaulay:

$$D = \frac{1}{mP}\left[1 \cdot \frac{c/m}{(1+y/m)^1} + 2 \cdot \frac{c/m}{(1+y/m)^2} + \cdots + m \cdot n \cdot \frac{100 + c/m}{(1+y/m)^{m \cdot n}}\right].$$

This formulas has three interpretations. First, simply by inspection, it is clear that D is the weighted average of coupon payment times. The weights are discounted coupon payments divided by the bond price. It is easy to verify that all weights sum up to one. Second, duration is an elasticity concept. The formula can be obtained by taking the derivative of the yield-to-maturity formula with respect to the yield, i.e.,

$$D = -\frac{dP}{dy} \cdot \frac{(1+y/m)}{P}.$$

Finally, duration is the location of the (gravity) center of the lever weighted by coupons:

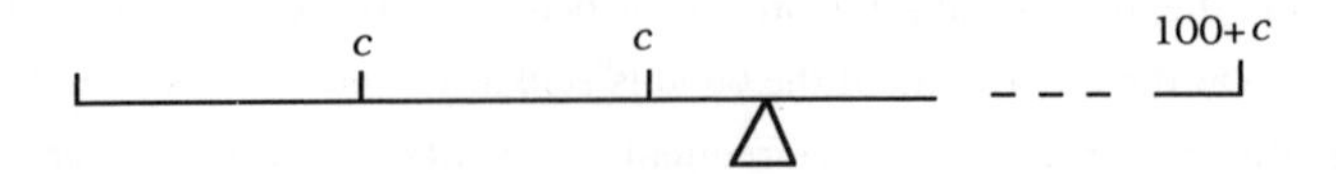

From this diagram, it is seen that the pure discount bond (no coupons) has the longest duration since the center needs to be right below the face value. The higher the coupon, the shorter the duration.

The definition of convexity is a measure of how duration changes. So it is the second derivative:

$$C = \frac{d^2P}{dy^2} \Big/ P$$

or expand it to get:

$$C = \frac{1}{m^2 P}\left[1 \cdot 2 \cdot \frac{c/m}{(1+y/m)^3} + 2 \cdot 3 \cdot \frac{c/m}{(1+y/m)^4} + \cdots + (mn)(mn+1) \cdot \frac{100 + c/m}{(1+y/m)^{m \cdot n+2}}\right].$$

There is no agreement on how convexity should be defined. Different books may have different formulas. The reason is that convexity has no simple interpretation (the unit of convexity is not clear) like duration and practitioners use convexity to get a better grasp of interest rate risk in different ways. As long as we catch the basic second order derivative, any definition is fine.

1.5.2 Immunization — The Use of Duration and Convexity

The use of duration is commonly known as immunization. It goes as follows. Suppose there is a liability due in 10 years. The amount is $215.06. Since the liability is certain, we can use a 10 year T Strip to find its present value, say $87.60. The yield calculated is approximately 9.4%. If a 20 year, 8% coupon bond has the same yield, then the price is also $87.60. So raising a 10 year debt is able to buy a 20 year bond. If we compute the duration of the bond, we find that it is 10 years which is also the duration of the debt.

Then we call this liability is immuned, immuned to interest rate risk or interest rate fluctuation. To see that, suppose that the interest rates change and the yield suddenly drops to 6%. Then the asset will be worth $114.72 but the reinvestment income at 6% for 10 years drops to $105.45 and the total is $220.17. This is enough to cover the liability of $215.06. If the interest rates rise so that the yield climbs to 12%, then the asset will be worth $77.40 but the reinvestment income will amount to $140.40 to a total of $217.80, also enough to pay off the debt. This gain on both sides makes the debt perfectly immuned for interest rate changes. However, as you might have seen, the change of interest rates is sudden and never happen again. The reinvestment income of the coupons for 10 years as well as the asset value both are calculated with the new yield. If there are more changes before the liability matures, immunization will face problems. The discussions of immunization have been fruitful. Interested readers can find interesting reading materials from the references at the end of the book.

The use of convexity is generally as a help to duration. In the above example, the convexity is higher for the asset than for the liability. Look at the following graph:

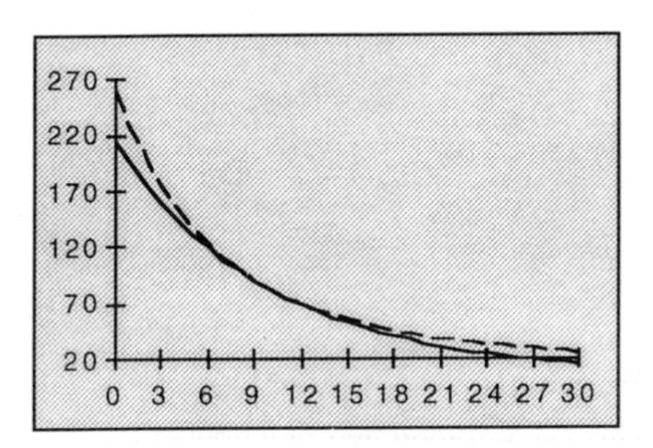

Figure 6. Duration/Convexity.

In the graph, the upper curve (dotted curve) is the asset and the lower curve (solid curve) is the liability. They tangent at 9.4%. Since the duration is measured by the slope, this tangent point is where we find the asset and the liability have the same value and the same duration. Since the asset is more curved, or has higher convexity, it gains in both directions. In bond management, we need to look for the highest convexity possible as our assets and lowest convexity possible as our liabilities. If the two are tangent, the portfolio is immunized.

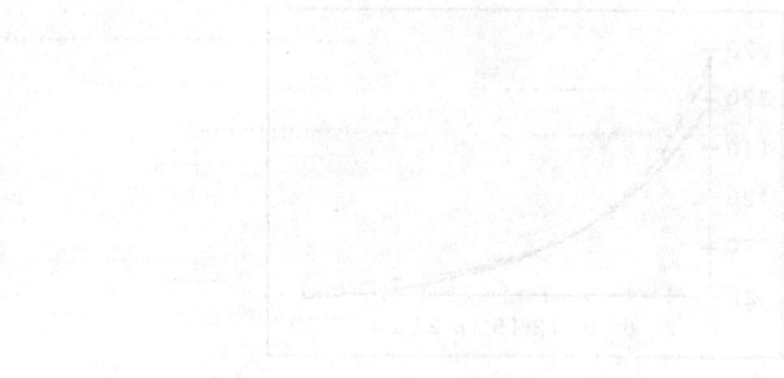

1/27/01

CHAPTER 2
TERM STRUCTURE MODELS

2.1 INTRODUCTION TO PRICED-BY-ARBITRAGE AND RISK-NEUTRAL PRICING

The arbitrage concept defined in Chapter 1 is the strongest definition. It needs no assumptions and no pricing formulas. As long as markets are liquid, it needs only traded assets and the price determined by this arbitrage is unique (or arbitrage profits will take place.) We name this strongest form of arbitrage "pure arbitrage". Sometimes such an arbitrage is not easy to be found. For example, American options have no pure arbitrage pricing to determine its price. The European option has an arbitrage pricing formula through put call parity but has no pure arbitrage pricing formula in terms of its underlying asset. Recall that for a pure arbitrage pricing formula to exist, a "complete market" is necessary, which requires enough related assets. For pricing a European call option by arbitrage, a bond, a put and its underlying asset are able to form a complete market but the underlying asset and the bond alone (no put) are not sufficient. For American options, since they can be exercised at any time and cash flows are uncertain, it is not possible to find enough assets to price them by pure arbitrage.

When pricing by pure arbitrage is not possible, we need to define a looser sense of arbitrage. Fischer Black and Myron Scholes (1973) found that in continuous time, all markets are complete if there are two different non-redundant assets and they price European stock options with only stocks and bonds. Since the arbitrage exists only in continuous time, the formula cannot be valid beyond infinite-decimal time interval. This is called continuous rebalancing. Duffie and Huang (1985) call it "dynamically complete markets". Now, all pricing models in the term structure literature are derived under continuous time and they are called arbitrage-free models in this loose sense. The risks certainly exist over a period of time since no one can trade continuously in reality. However, given that is no better solution and the continuous time model has been used quite successfully, we will treat it as it were a true arbitrage-free model.[2]

If an asset cannot be priced by arbitrage, either in pure sense or in loose sense, we can always identify a utility-based formula. Actually, there have been a number of utility-based formulas for equity options. We know that a reason for having a utility-based formula is because there exists no arbitrage in discrete time. Even in continuous time, sometimes there will be utility-based formulas. These incidents occur because

[2]It should be noted that binomial models, although discrete, share the identical assumption with the continuous time models.

some options (or derivative contracts) are written on non-traded assets, for example, interest rate options. The intuition is quite simple. If the underlying asset is not traded, then even in continuous time, the market cannot be complete, then there of course exists no arbitrage pricing.

A utility-based model is more cumbersome because it adopts a quite unreasonable assumption, which is that the economy permits one person (or all people in this economy are identical) and his risk attitude has to follow some fixed structure over time. Although there can be some modifications to this stringent assumption, any more realistic modification will complicate the solution very badly. In this chapter, we shall discuss term structure models which serve as a base to the pricing of all interest rate derivative contracts. All models introduced in this chapter are based upon an instantaneous rate which is unobservable and non-traded. And the models are all utility-dependent in which a single representative agent represents the economy.

In the previous chapter, we learn that we can price a pure discount bond with a series of coupon bonds and the pricing is an arbitrage-free one. If we do not want to price pure discount bonds with coupon bonds but with a fundamental state variable, then it of course would need no arbitrage pricing. But we can still identify a formula. Since there is no arbitrage pricing, the formula will be utility-dependent. Given that there is no nice pricing formula by arbitrage, the formula can be developed in continuous time as well as in discrete time. However, as a matter of convenience, we show continuous time results in this chapter since all other results are all defined in continuous time. It can be seen later that, although the pure discount bonds themselves have no pure arbitrage solutions, all other derivatives do (in continuous time).

It should be noted that equilibrium formulas should also disallow arbitrage opportunities, or formulas are not in equilibrium. A lot of authors call models developed under a utility function "equilibrium models" and the other models "arbitrage-free models" if they are derived using the Black–Scholes methodology. This is very misleading because "equilibrium models" should be arbitrage free and "arbitrage-free" models may or may not have arbitrage opportunities. Whether or not a model needs a utility function depends not on how the model is derived but on if the markets are (dynamically) complete and continuous hedging is possible.[3] If not, then although a model is derived with the Black–Scholes methodology, it may still contain arbitrage opportunities due to erroneously defined hedging process or risk premium.

Before we show the models, we first need to examine the pure expectation

[3]Some processes have no diffusion. Then these processes will not be able to produce utility-free results.

Martingales & Arbitrage in Multi-period Securities Mkts
JET 1979 p 381-408

hypothesis in the term structure theory. If the short term rate moves as follows:

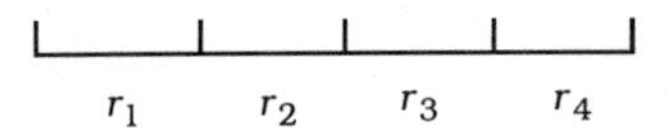

a $1 face value pure discount bond should then be priced at:

$$P = \frac{1}{(1+r_1)(1+r_2)(1+r_3)(1+r_4)}.$$

Given that r_2, r_3 and r_4 are unknown, the price is also unknown. The pure expectation hypothesis says that the price should be a simple expectation of this uncertain quantity, i.e.,[4]

$$P = \hat{E}\left[\frac{1}{(1+r_1)(1+r_2)(1+r_3)(1+r_4)}\right].$$

The hat on top of the expectation operator is to show the expected value is computed as if the representative agent were risk-neutral. This is known as "risk-neutral" pricing. The basic theory is explained by Harrison and Kreps (1979). It should be noted that the risk-neutral expectation does not mean the agent is risk-neutral. It is merely a pricing gimmick that makes the computation of the expectation easier. If the agent is risk-averse, we will take his risk premium out of the economy so that the agent is risk-neutral in this "equivalent" economy. Since a risk-neutral utility function is linear in wealth (or consumption), the computation of the expectation is easy. Because the terminology "risk-neutral" refers to the equivalent economy not to the actual risk attitude of the agent, some researchers prefer call the expectation "risk-neutralization" rather than "risk-neutral".

The concept of risk-neutral pricing is quite easy to understand. Suppose we have a binomial economy where the only asset in the economy has only two future values with probabilities of p and $1-p$ of up and down respectively:

[4]Ingersoll shows (1989) shows that there are several forms of expectation hypothesis and there is only one, pure expectation hypothesis, that is consistent with no arbitrage assumption.

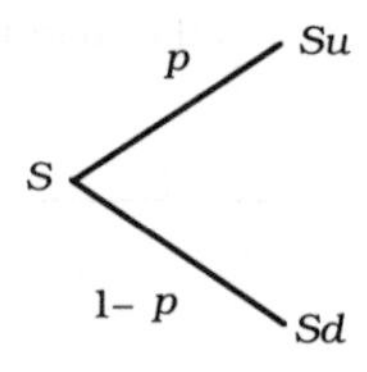

The expected value today of future prices is certainly:

$$E[S] = p \times Su + (1 - p)Sd.$$

The expected percentage return is a risky one and should be higher than the risk free rate since it carries an uncertainty risk:

$$k = \frac{E[S]}{S} > R$$

where R is 1 plus the risk free return. The reason that k is greater than R is that a risk-averse person will ask for compensation for taking risks. If this person is, say for some reason, risk loving, then his k will be actually lower than R because he is so happy taking risks that he is willing to give up some return for being able to take risks. Finally, if he is risk-neutral, then he is insensitive to risks. High or low risks to him means exactly the same and therefore, he will always ask the same return. And by the above analyses we know he will ask for exactly the risk free return. His k is equal to R. Risk aversion is always assumed by all economics theories. However, in pricing, we find risk aversion assumption a very annoying one because pricing formulas sometimes are not easy to derive. Because the utility function of a risk-neutral person is linear in wealth, the mathematics is so much easier. But how can we use the risk-neutral methodology while individuals in the economy are assumed to be risk-averse? Thanks to the recent developments in option pricing, this is possible now.

Suppose we create a risk-neutral person (keep in mind that this person never exists in our economy) by solving the following equation for his view of probabilities:

$$\begin{aligned} S &= \hat{E}[S] / R \\ &= \frac{\hat{p}Su + (1 - \hat{p})Sd}{R} \cdot \end{aligned}$$

The reason is that he expects exactly the risk free return and therefore we must know his probabilities by equating his expectation of the stock price to $S \times R$. Since u and d are known, the risk-neutral probability can be solved. Now, suppose there is riskier asset traded in our economy, say option. Since option has more extreme payoff, it is riskier than the stock and the risk-averse person will ask for more return. The expected value of the option by the risk-averse person is:

$$E[C] = pCu + (1-p)Cd$$

and this return should be more than that on the stock to justify the extra risk:

$$k_C = \frac{E[C]}{C} > \frac{E[S]}{S} = k$$

So, to obtain today's option price, we need to know the required return of the risk-averse person and of course this required rate of return is a function of the degree of risk aversion. But look at this hypothetical risk-neutral person. We know his expected option value is:

$$E[C] = \hat{p}Cu + (1-\hat{p})Cd = R \times C.$$

This expected value must equal $R \times C$ because he requires only risk free return regardless of the risk of the asset. We know that we can solve for $\hat{p}$ from the stock market. Now plug it in the option market and we can calculate today's option price. This risk-neutral person has never existed but by imagining him we can make a very difficult problem (risk aversion dependent) a very easy one. The probability, $\hat{p}$, is called the "risk-neutral" probability or "pseudo" probability. Once this risk-neutral probability becomes known, we can calculate the expected value of any asset and then discount it at the risk free rate. This methodology is first called risk-neutral pricing and later understood as priced by arbitrage. The difference between real probability p and the risk-neutral probability $\hat{p}$ is called risk-neutralization or change of probability space.

[EXAMPLE 1]

Suppose the current stock price is $50 and it is expected with some (true) probability to go up to $55 or down to $45. The risk-free rate is 10%. If there is a call option maturing at the next period with a strike price of

When markets are incomplete pricing by arbitrage is meaningless

$50. Then what is the risk-neutral probability and what is the current option price?

$$\hat{p} = \frac{R-d}{u-d} = 0.75 ,$$

Plugging into the risk-neutral expectation to find the call price:

$$C = \frac{1}{R}\hat{E}[C] = \frac{0.75 \times 5 + 0.25 \times 0}{1.1} = 3.5$$

Note that for the risk-neutral pricing methodology to work, the existence of the stock market is necessary, otherwise we cannot solve for risk-neutral probability $\hat{p}$. This requirement is known as the assumption of the complete markets. If the stock market did not exist, then there would be no current stock price for us to calculate $\hat{p}$ and the risk-neutral pricing methodology would not work. In this case, we must assume a current stock price in order to be able to solve for $\hat{p}$. But note that once we assume a current stock price, we are at the same time assuming a value for p, which means we are imposing unknowingly a specific degree of risk aversion of our economy. This specific of degree of risk aversion can be directly used in figuring out the required rate of return of the option, k_C, and there is no need for the risk-neutral person to exist. Then the pricing problem is back to the basic asset pricing problem. In other words, when markets are incomplete, priced by arbitrage or risk-neutral pricing is meaningless.

However, nowadays, many people assume a fixed difference between p and $\hat{p}$ and continue to use the risk-neutral pricing methodology to figure out contingent claim prices when their underlying assets are not traded. Simply speaking, they assume a current "equilibrium" price for the stock and the risk-averse probability p, use it to figure out $\hat{p}$ by further assuming a fixed "risk premium" between p and $\hat{p}$. And then use $\hat{p}$ to calculate all claim prices. From our example above, we know that this is very dangerous. The option prices from this possibly wrong $\hat{p}$ and from directly using k_C could be different. Cox, Ingersoll, and Ross (1985a) and Ingersoll (1987) although not clear both make a point of this issue.

Now back to our term structure discussion. Once we find the risk-neutral probabilities, we can simply find all asset prices by discounting its risk-neutral expected value. For the discount bond price, we know that the instantaneous short rate is not a traded asset, there is no market for it. As a result, the incomplete markets argument will come into play. In this case, we need to explicitly assume a utility function to obtain the risk aversion, or more precisely, the risk premium between the actual probability space and the risk-neutral probability space. As emphasized above, this risk premium needs to be derived from the utility function; it cannot be assumed of any form. For the models discussed in this chapter, they are fine. Making the face value of a unit bond \$1, in continuous time, the pricing formula becomes:

$$P(t,T) = \hat{E}_t\left[\exp\left(-\int_t^T r(u)du\right)\right].$$

It is noted that the bond price is symbolized by two time points, t and T, which are beginning and ending times of the integral which also resemble the current time and the maturity time of the bond. The subscript t of the expectation operator represents that the expectation is taken at time t.

If the (random) movement of future interest rates over time were assumed to follow some known distribution, then computing the bond price would be as simple as carrying out an expected value (risk-neutral). In this chapter, we shall give two explicit examples of the process of future interest rates where easy solutions for the bond price can be identified. In the literature, there have been many models developed but only two are widely cited and studied. They are the Vasicek model developed in 1977 and the Cox–Ingersoll–Ross model that was developed at the same time but not published until 1985. We shall discuss them in details to raise the basic pricing issues while we shall also review other popular models.

2.2 SINGLE-FACTOR MODELS

The early work on the term structure gives us great insights of how the term structure of interest rates can be or should be modeled. Later on due to problems of these early models, more recent models have been developed. In this section, we provide basic pricing methodologies and insights through these early models and then in the next two sections we present all various modern models. We pay, specifically, a lot of attention to the Vasicek model (1977) and the Cox–Ingersoll–Ross model (1985b).

2.2.1 Vasicek Model

Since the interest rate, $r(t)$, randomly moves over time, we in general use a random process to model it. A continuous time process, or often called diffusion process, is generally used to model r. In this section and the next section, the processes have differential representations which are used in finance literature to describe these processes. In the Vasicek model (1977), the interest rate process looks as follows:

$$dr = \alpha(\mu - r)dt + \sigma dW$$

where α, μ, and σ are fixed constants and $W(t)$ is a standard Wiener process. For people who are not familiar with the Wiener process, they can view dW as normal variate with mean 0 and variance dt. This differential equation belongs to a series of equations called "linear stochastic differential equations" and has a nice distribution for r. At any given time the distribution for r is normal with mean and variance, and covariance as follows:

$$\begin{cases} E_t[r(s)] = r(t)e^{-\alpha(s-t)} + \mu\left(1 - e^{-\alpha(s-t)}\right) \\ \text{cov}_t[r(u), r(s)] = \frac{\sigma^2}{2\alpha} e^{-\alpha(s+u-2t)}\left(e^{2\alpha(u-t)} - 1\right) \quad \text{for } u < s \end{cases}.$$

The conditional variance is found by solving:

$$\begin{aligned} \text{var}_t[r(s)] &= \text{cov}_t[r(s), r(s)] \\ &= \frac{\sigma^2\left(1-e^{-2\alpha(s-t)}\right)}{2\alpha} \end{aligned}.$$

Other than that the distribution is known, the process also has a nice feature that r, the short rate is "mean-reverting". If r is larger than μ at any time, then μ–r is negative, which means in the next instant (dt later), the interest rate will likely to fall because expected change of the interest rates:

$$E[dr] = \alpha(\mu - r)dt + \sigma E[dW] = \alpha(\mu - r)dt < 0$$

is negative. On the other hand, if r is smaller than μ, then the expected change is positive and the interest rate in the next instant will rise. As a result, the interest rate will go around the level μ. The parameter σ simply provides the magnitude of uncertainty.

To find the solution to the pure discount bond, we need to compute explicitly the expectation. To do that, it is convenient to find the distribution of the integral of r, i.e., we shall first find the distribution of:

$$R \equiv \int_t^T r(u)du.$$

Since r is normally distributed, it follows naturally that R is also normally distributed with mean and variance as:

$$\begin{aligned} E_t[R] &= \int_t^T E_t[r(s)]ds \\ &= r(t)\left(\frac{1-e^{-\alpha(T-t)}}{\alpha}\right) + \mu\left[T - t - \left(\frac{1-e^{-\alpha(T-t)}}{\alpha}\right)\right] \end{aligned}$$

and

$$\begin{aligned} V[R] = \operatorname{cov}[R, R] &= \int_t^T \int_t^u 2\operatorname{cov}_t[r(u), r(s)]\, du\, ds \quad (\forall u < s) \\ &= 2\int_t^T \int_t^u \frac{\sigma^2}{2\alpha} e^{-\alpha(s+u-2t)}\left(e^{2\alpha(u-t)} - 1\right) du\, ds \\ &= \frac{\sigma^2}{\alpha^2}\left[T - t - \frac{e^{-2\alpha(T-t)}}{2\alpha} + 2\frac{e^{-\alpha(T-t)}}{\alpha} - \frac{3}{2\alpha}\right]. \end{aligned}$$

It is then clear that the bond solution is a moment generating function of the normal variable R. Before the final solution can be reached, we should note that the expectation is a risk-neutral one, not the one above. We learn from the previous section that once a risk-neutral process (probabilities) can be identified, pricing becomes easy because all assets then earn only the risk-free return regardless of their actual risks. Obtaining a risk-neutral distribution requires solving a general equilibrium economy and is beyond the scope of this book. Here, we shall take the result without proof that the risk-neutral mean and variance are as follows:

$$\begin{aligned} \hat{E}_t[R] &= \int_t^T \hat{E}_t[r(s)]ds \\ &= r(t)\left(\frac{1-e^{-\alpha(T-t)}}{\alpha}\right) + \left(\mu - \frac{q\sigma}{\alpha}\right)\left[T - t - \left(\frac{1-e^{-\alpha(T-t)}}{\alpha}\right)\right] \end{aligned}$$

and

$$\hat{V}_t[R] = V_t[R].$$

It is seen here that the risk-neutral variance remains unchanged and the risk-neutral mean is slightly changed by a risk parameter, q which is fixed under log utility.[5] The details can be found in a number of places including Cox, Ingersoll, and Ross (1985a), Vasicek (1977), and Ingersoll (1989). The fact that the variance is unchanged is a specific result tied to normality. We shall see in the next model where the variance will change.

The solution can be written out explicitly as:

[5]The derivation is given by Campbell (1986).

$$P(t,T) = e^{-\hat{E}_t[R] + \frac{\hat{V}_t[R]}{2}}$$
$$= e^{-r(t)F(t,T) - G(t,T)}$$

where

$$F(t,T) = \frac{1 - e^{-\alpha(T-t)}}{\alpha}$$
$$G(t,T) = \left(\mu - \frac{q\sigma}{\alpha} - \frac{\sigma^2}{2\alpha^2}\right)[T - t - F(t,T)] + \frac{\sigma^2 F^2(t,T)}{4\alpha}.$$

Given 4 parameters, α, μ, σ, and q, we can easily calculate the bond price today for any given maturity.

[EXAMPLE 2]

Given the estimates from Chen and Yang (1995) who estimate parameters using interest rate data from January 1988 to June 1993:

$$\alpha = 0.2456$$
$$\mu = 0.0648$$
$$\sigma = 0.0289$$
$$q = -0.2718$$
$$r = 0.0600$$

we can calculate 30 bond prices from one year to maturity to 30 years to maturity and also draw the yield curve. The thirty prices and the their yields are presented as follows.

time to maturity	bond price	yield to maturity	time to maturity	bond price	yield to maturity
1	0.9380	6.4053%	16	0.2639	8.3257%
2	0.8740	6.7328%	17	0.2413	8.3622%
3	0.8106	7.0008%	18	0.2207	8.3951%
4	0.7491	7.2224%	19	0.2017	8.4249%
5	0.6905	7.4075%	20	0.1844	8.4520%
6	0.6352	7.5635%	21	0.1686	8.4767%
7	0.5835	7.6960%	22	0.1541	8.4993%
8	0.5354	7.8095%	23	0.1409	8.5201%
9	0.4908	7.9074%	24	0.1288	8.5392%
10	0.4497	7.9923%	25	0.1177	8.5568%
11	0.4118	8.0664%	26	0.1076	8.5732%
12	0.3769	8.1314%	27	0.0984	8.5883%
13	0.3449	8.1889%	28	0.0899	8.6024%
14	0.3155	8.2398%	29	0.0822	8.6156%
15	0.2886	8.2851%	30	0.0751	8.6279%

To graph the last column, we obtain what is commonly used the yield curve. This yield curve is also called the discount curve by a lot of people because it represents the term structure of interest rates for only discount bonds.

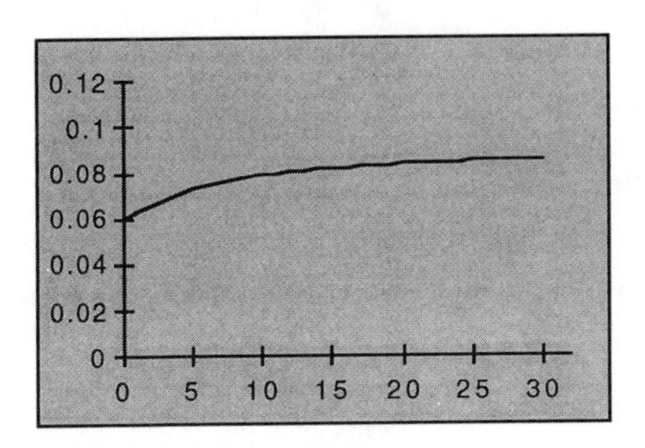

2.2.2 Cox–Ingersoll–Ross Model

We present the CIR model in a similar manner. The only difference between the CIR model and the Vasicek model is the assumption of the interest rate process. CIR have the following assumption of the interest rate differentials:

$$dr = \kappa(\theta - r)dt + \upsilon\sqrt{r}dW.$$

vs. Vasicek $dr = \alpha(\mu - r)dt + \sigma dW$

The drift part is identical to the Vasicek assumption. The difference lies in the diffusion part in which the CIR assumption has a square root of r in the process. This tiny difference makes the model completely different. This process is much more difficult to deal with but it has an advantage that all future interest rates will be strictly non-negative. This is a very desirable feature for nominal interest rates. Because if interest rates can be negative, then bond prices can exceed one. This is a clear violation of the arbitrage rule since anyone can short the bond that is sold more than \$1 and need to pay back for only \$1 in the future.

Since the addition is a "square root" of r, it is called the "square root" process of the interest rate. The non-linearity in the equation changes the distribution of the interest rate. It is no longer normally distributed. The distribution has been worked out by Feller (1951) to be a scaled non-central chi-squared distribution. The distribution looks like: Two Singular Diffusion Problems Annals of Math July, 1951. Vol 54 No 1.

$$f(r(s)|r(t)) = ce^{-c(r(s)+\xi)}\left(\frac{r(s)}{\xi}\right)^{d/2} I_d\left(2c\sqrt{\xi r(s)}\right)$$

where

$$d = \frac{2\kappa\theta}{\upsilon^2} - 1$$
$$c = \frac{2\kappa}{\upsilon^2\left(1-e^{\kappa(s-t)}\right)}$$
$$\xi = e^{-\kappa(s-t)}r(t)$$

$I_d(\cdot)$ is the modified Bessel function of the first kind of order d.

This density function is not a non-central chi-squared distribution. But if we set

$$x = 2cr,$$

then x is a non-central chi-squared random variable which has the following density function:

$$f(x(s)|x(t)) = \frac{1}{2}e^{-c(x(s)+\Lambda)}\left(\frac{x(s)}{\Lambda}\right)^{d/2} I_d\left(\sqrt{\Lambda x(s)}\right)$$

where $\Lambda = 2c\xi$ is the degree of non-centrality and $2(d+1)$ is the degree of freedom. The properties of the non-central chi-square distribution can be found in Johnson and Kotz (1970). As in the Vasicek model, we need to identify the risk-neutral distribution for the interest rate. By CIR (1985b), the risk-neutral distribution for this model under log utiilty is to replace κ by $\kappa+\lambda$:

$$\hat{f}(r(s)|r(t)) = \hat{c}e^{-\hat{c}(r(s)+\hat{\xi})}\left(\frac{r(s)}{\hat{\xi}}\right)^{d/2} I_d\left(2\hat{c}\sqrt{\hat{\xi} r(s)}\right)$$

where

$$\hat{c} = \frac{2(\kappa+\lambda)}{\upsilon^2\left(1-e^{(\kappa+\lambda)(s-t)}\right)}$$
$$\hat{\xi} = e^{-(\kappa+\lambda)(s-t)}r(t).$$

In this risk-neutral distribution, the degree of freedom remains unchanged but the degree of non-centrality becomes $\hat{\Lambda} = 2\hat{c}\hat{\xi}$. Then the bond price is to solve the following integral:

$$P(t,T) = \hat{E}_t\left[\exp\left(-\int_t^T r(u)du\right)\right]$$
$$= \int_0^\infty e^{-R}\hat{f}(r(T)|r(t))dr(T).$$

Unlike the normality case, the distribution of R is not known. The only way to obtain the distribution of R, is to solve the Laplace transform, which leads to solving a partial differential equation. The technique of solving the Laplace transform or the partial differential equation is beyond the scope of this book. We shall provide the answer without proof as follows:

$$P(t,T) = A(t,T)e^{-r(t)B(t,T)}$$

where

$$A(t,T) = \left(\frac{2\gamma e^{(\kappa+\lambda+\gamma)(T-t)/2}}{(\kappa+\lambda+\gamma)(e^{\gamma(T-t)}-1)+2\gamma}\right)^{2\kappa\theta/\upsilon^2},$$
$$B(t,T) = \frac{2(e^{\gamma(T-t)}-1)}{(\kappa+\lambda+\gamma)(e^{\gamma(T-t)}-1)+2\gamma}, \text{ and}$$
$$\gamma = \sqrt{(\kappa+\lambda)^2 + 2\upsilon^2}.$$

The parameter λ serves the same purpose as q in the Vasicek model as a risk parameter that is implied by the single agent's utility function. Likewise, it is constant under log utility. Similar to the Vasicek model, this model can be easily used to find all bond prices and also the yield curve. With the same data and methodology, Chen and Yang also estimate the CIR model as follows.

[EXAMPLE 3]

The parameters are:

κ=0.2456

θ=0.0648

υ=0.1499

λ=-0.1290

r=0.0600

we can calculate 30 bond prices from one year to maturity to 30 years to maturity and also draw the yield curve. The thirty prices are in the following table.

time to maturity	bond price	yield to maturity	time to maturity	bond price	yield to maturity
1	0.9379	6.4079%	16	0.2642	8.3201%
2	0.8738	6.7464%	17	0.2418	8.3516%
3	0.8100	7.0260%	18	0.2213	8.3799%
4	0.7481	7.2567%	19	0.2025	8.4054%
5	0.6891	7.4473%	20	0.1853	8.4285%
6	0.6336	7.6054%	21	0.1696	8.4495%
7	0.5818	7.7372%	22	0.1552	8.4687%
8	0.5338	7.8478%	23	0.1420	8.4862%
9	0.4893	7.9413%	24	0.1300	8.5024%
10	0.4484	8.0209%	25	0.1189	8.5172%
11	0.4107	8.0893%	26	0.1088	8.5310%
12	0.3761	8.1484%	27	0.0996	8.5437%
13	0.3444	8.1998%	28	0.0911	8.5556%
14	0.3153	8.2450%	29	0.0834	8.5666%
15	0.2886	8.2848%	30	0.0763	8.5770%

The graph is presented as:

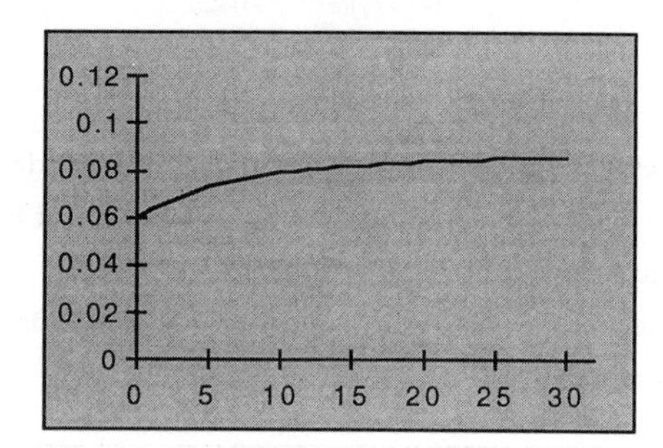

2.2.3 Dothan Model

Another interesting model with non-negative interest rates, although not widely used, is the Dothan model (1978). The Dothan model differs in that the interest rate r follows a log normal distribution. This is the same log normal model as the Black and Scholes model for the stock option except that there is no drift term:

$$dr = \eta r dW .$$

It is seen that this model does not allow mean reversion. This model is also called the geometric random walk or elastic random walk model. Like the CIR model, the distribution of R is not known. Furthermore, there is no simple solution for the Laplace transform. As a result, the bond solution is not easy and the computation is complex. Readers can refer to Dothan's original work for the formula.

2.2.4 Constantinides Model

Constantinides (1992), in view of the conflict between simplicity of a model and non-negativity of interest rates, came up with a solution by avoiding modeling r directly. He, by modeling the "pricing kernel" with a normal process, provides a one-factor model where all future interest rates are non-negative. His model is able to prevent negative interest rates and produce simple solutions to the term structure and its contingent claims at the same time. The drawback of the model is that the option formula depends on an unknown and unobservable factor. You have always to give up something if you acquire another. We have seen that we give up simplicity to acquire non-negativity or keep both to give up observability.

Instead of using the standard risk-neutral pricing methodology, Constantinides' model uses directly the stochastic Euler equation which is defined in the original probability space:

$$E_t\left[\frac{X(T)M(T)}{X(t)M(t)}\right] = 1$$

where X can be any asset. $M(t)$ can be interpreted as the marginal utility at time t. It is known that the ration of the two marginal utilities is called the marginal rate of substitution which can be regarded as a stochastic discount factor in the original probability space. With risk neutralization, this discount factor can be represented as discounting at the risk-free rate (stochastic):

$$X(t) = \hat{E}_t[X(T) / R(t,T)].$$

This then returns to the usual risk-neutral pricing methodology. However, in the Constantinides model, there is no need to perform the transformation because the model is directly defined on the original space.[6] The discount bond price is then a straightforward expectation of the marginal rate of substitution:

[6] Intuitively, we can view the Constantinides model as a model on the risk-adjusted rate; therefore there is no need to perform the transformation. Note that the need for the transformation is because we want to avoid the complexity of determining the risk-adjusted rates for various instruments.

Ho Lee model uncertainty by putting perturbation fns on the forward prices.

$$P(t,T) = E_t[M(T)] / M(t).$$

The distribution assumption of $M(t)$ and the closed-form solution for the bond price can be found in the original paper.

There is one more advantage of the Constantinides model. Since the short rate is not a linear function of the state variable, the yields of different-maturity bonds are not perfectly correlated. This is a certain improvement on the Vasicek/CIR models.

2.2.5 Discussions

Many studies have shown that the one-factor models shown above are not suitable in modeling the current term structure.[7] Sine then, the research has been on developing a term structure model that can well describe the curvature of the observed term structure. One approach is to take the current term structure as given and incorporate term structure fluctuations in various ways. Ho and Lee (1986) model the uncertainty by putting perturbation functions on the forward prices. Black, Derman, and Toy (1990), or BDT, model is similar to the Ho–Lee model except that the distribution of the short rate is log normal. The BDT model is richer than the Ho–Lee model in that it fits also the volatility curve. The continuous time models by Heath, Jarrow, and Morton (1992) and Hull and White (1990) let the parameters in the stochastic processes of the instantaneous rate be deterministic functions of time. All of these models are considered "time-dependent" models. These models can fit the yield curve but they do not have an easy form for the bond price.

Yet another approach is to add more factors to the term structure model. Richard (1978), Brennan and Schwartz (1979), Langetieg (1980), Cox, Ingersoll, and Ross (1985), and Longstaff and Schwartz (1992) have developed two-factor term structure models where either an arbitrage-free or utility-based methodology is used. In the Richard and Brennan–Schwartz models, factors are chosen arbitrarily. There is little theoretical support to such choices. Cox, Ingersoll, and Ross, on the other hand, construct a model under a general equilibrium framework in which they derive factors and their dynamics from production technology and an additive utility function of a representative agent. They find that in equilibrium, the instantaneous rate can be

[7] The empirical studies that show one-factor models can not fit the yield curve are Chen and Scott (1993) and Pearson and Sun (1990) for maximum likelihood; Heston (1989) and Gibbons and Ramaswamy (1993) for generalized method of moments; Litterman and Scheinkman (1991) for factor analysis.

expressed by separate factors. There are two ways of constructing two factors in their model. The first method is to directly decompose the instantaneous rate into two factors, each of which follows some stochastic process. This decomposition produces a simple solution of the term structure. As pointed out by Chen and Scott (1993), this method is tractable empirically and maximum likelihood estimation is possible. Another method is to observe that the CIR model can be interpreted as a stochastic volatility model because the volatility of the instantaneous interest rate is a function of the two factors. Longstaff and Schwartz (1992) adopt this approach and derive their two-factor term structure model. The two models are essentially the same. The difference is merely a choice of simplicity or factor observability. In the Chen–Scott model, factors are not observable while in the Longstaff–Schwartz model, the mathematical tractability is lost.

2.3 MULTI-FACTOR MODELS

The disability of fitting the yield curve raises the need for multiple factors. The development of the multi-factor models can be divided into two stages. At an earlier stage, factors are arbitrarily specified. Brennan and Schwartz (1978) use short and long rates as factors. Richard (1978) uses the real interest rate and inflation as two factors.

2.3.1 Brennan–Schwartz Model

Brennan and Schwartz (1978) argue that for the yield curve to be explained, at least two factors are necessary. And these two factors need to be the two ends of the yield curve, i.e., short and long rates. The short rate has mean reversion to the long rate and follows a log normal process while the long rate follows another log normal process. To be specific:

$$
\begin{aligned}
d\ln r &= a(\ln \ell - \ln r)dt + b_1 dW_1 \\
d\ell &= \ell a(r, \ell, b_2)dt + b_2 \ell dW_2
\end{aligned}
$$

where $E[dW_1 dW_2] = \rho dt$. Obviously, this problem has no closed form solution to the bond price. One has to pursue numerical solutions for the answer. The partial differential equation can be derived through the standard arbitrage argument as follows:

$$\frac{1}{2}P_{11}b_1^2r^2 + P_{12}\rho b_1 r b_2 \ell + \frac{1}{2}P_{22}b_2^2\ell^2 + P_1(\ln \ell - \ln r - \lambda_1 b_1 r) + P_2(\ell a_2 - \lambda_2 b_2 \ell) - P_t = rP$$

Brennan and Schwartz use the finite difference method to solve this problem.

2.3.2 Richard Model

Richard's two factors are the real rate of interest, ρ, and the rate of inflation, π, which are independent of each other and follow the following square root processes:

$$\begin{aligned} d\rho &= a_\rho(\rho - \rho^*)dt + b_\rho\sqrt{\rho}dW_1 \\ d\pi &= a_\pi(\pi - \pi^*)dt + b_\pi\sqrt{\pi}dW_2 \end{aligned}.$$

Richard then finds that the nominal rate of interest is more than the sum of real rate and inflation:

$$r = \rho + \pi(1 - \mathrm{var}[dP / P])$$

where P represents the price level whose expected change is inflation. Following the same analysis, the solution to the nominal bond price is:

$$\begin{aligned} P(t,T) &= \hat{E}_t\left[\exp\left(-\int_t^T r(u)du\right)\right] \\ &= \hat{E}_t\left[\exp\left(-\int_t^T \rho(u)du\right)\right]\hat{E}_t\left[\exp\left(-\int_t^T \pi(u)du\right)\right]\exp(1 - V_P) \end{aligned}.$$

Note that this solution for the bond is a product of the two CIR results with an additional term $1-V_P$. The mode has two disadvantages. First, the specification of the two state variables (real rate and inflation) is ad hoc. Second, the independent assumption between the real rate and the inflation is counter empirical. Empirically, researchers have found negative relation between the real rate and inflation. It is noted in Chen and Scott (1992) that two correlated square root processes will lead to intractable solution for the bond price.

Richard's model would have a nice solution if the two correlated processes are allowed normal. In Chen (1995), the bond solution with two correlated normal process is derived in closed form. Therefore, taking Chen's solution and multiplying it by $1-V_P$ gives the solution to the bond price and the correlation can be negative.

2.3.4 Cox-Ingersoll-Ross/Langetieg Model

Following the suggestion by CIR, the instantaneous rate, r, is decomposed into two independent factors, y_1 and y_2 (i.e., $r = y_1 + y_2$). Then the bond solution is:

$$\begin{aligned} P(t,T) &= \hat{E}_t\left[\exp\left(-\int_t^T r(u)du\right)\right] \\ &= \hat{E}_t\left[\exp\left(-\int_t^T y_1(u)du\right)\right]\hat{E}_t\left[\exp\left(-\int_t^T y_2(u)du\right)\right]. \\ &= P_1(t,T)P_2(t,T). \end{aligned}$$

If each of the factors follows a Vasicek assumption, then each of the P's will be a single-factor solution described in the previous section. If the factors take CIR assumptions, then the bond price will be the product of the two CIR formulas.

This formulation is simple. It has nice solution like the single-factor model and remains under the general equilibrium framework. The drawback is that the two factors are unobservable so it is difficult to identify them. As a result, it is difficult for them to carry economic meanings. However, as Chen and Scott (1993) point out, the problem can be alleviated by running a correlation of the factors with known interest rates so that factors can be understood of their economic meanings. We shall be more specific when we discussion the empirical part of the models.

2.3.5 Longstaff-Schwartz Model

The Longstaff and Schwartz model is of little difference from the CIR model mentioned earlier. They differ in that they map unobservable factors into observable ones which are believed to be the most important factors in determining the term structure of interest rates. Using their notation, we write the two state variables as:

$$\begin{aligned} dy_1 &= (a - by_1)dt + c\sqrt{y_1}dW_1 \\ dy_2 &= (d - ey_2)dt + f\sqrt{y_2}dW_2 \end{aligned}$$

where $dW_1 dW_2 = 0$. By CIR, the equilibrium rate of interest and its volatility are:

$$\begin{aligned} r &= \alpha y_1 + \beta y_2 \\ V &= \alpha^2 y_1 + \beta^2 y_2 \end{aligned}$$

Any miss in the fitting of the yield curve can cause significant consequences in pricing derivatives.

Solving these simultaneous equations and using Ito's lemma, we can derive the dynamics for dr and dV. The closed-form solution for the bond price therefore comes with no surprise. The difference between this closed form solution and the previously mentioned closed-form CIR model is that the short rate is a weighted sum of the two factors while in the previous CIR model the short rate is a simple sum of the two factors. Note that this changes the distribution of the short rate. However, the changes are insignificant. As a result, the two models can be treated as identical models.

The advantage of the Longstaff and Schwartz model is of course that factors are observable, so parameters can be directly estimated from data. When factors are unobservable, they need to be computed themselves which adds complexity to the estimation. However, since the process assumption is imposed directly on factors, maximum likelihood estimation is possible. In their paper, Longstaff and Schwartz need to employ a less powerful GMM method for parameter estimation.

2.4 TIME-DEPENDENT PARAMETER MODELS

p43 $dr = \theta(t)dt + \sigma d\hat{W}$

2.4.1 Ho–Lee Model

At first, when Ho and Lee first discovered their model, it was not interpreted as a time-dependent model. It was not until Dybvig (1989) that people have come to realize that the Ho–Lee model is a time-dependent parameter term structure model and it can be presented as a closed form solution similar to Vasicek. We shall discuss Dybvig's interpretation later. First, we shall talk about the model in its original form.

Ho and Lee realize that a model would be useless if it cannot fit the yield curve. Although multi-factor models can improve the fitting, nonetheless, they still cannot fit well enough the yield curve. Moreover, any miss in the fitting of the yield curve can cause significant consequences in pricing derivatives. What they have come up with is a solution that, instead of pricing bonds, takes bond prices as given. Since their model takes bond prices as given, their model cannot be used to find bond prices. This is a dramatically different approach than what have been discussed above. The Ho–Lee model is used for interest rate derivatives, such as options.

They start from the observed current term structure and calculate a series of forward prices. The binomial evolution of the forward prices is done by imposing perturbation functions. Formally, the term structure of pure discount bonds is defined as:

$$P(0,1),\ P(0,2),\ \cdots,\ P(0,n).$$

With certainty, we know that the one-year bond price one-year from now should equal today's one-year forward price on a one-year bond, or:

$$P(1,2) = \frac{P(0,2)}{P(0,1)} = \Psi(0,1,2).$$

To incorporate uncertainty, a binomial tree is created by adding perturbations for up and down states:

$$P(1,1,2) = \frac{P(0,2)}{P(0,1)}u(1)$$
$$P(0,1,2) = \frac{P(0,2)}{P(0,1)}d(1).$$

The newly added index (first argument) represents the state; 1 for up and 0 for down. With this logic, we can create two yield curves in the next period as:

$$P(1,1,i) = \frac{P(0,i)}{P(0,1)}u(i-1)$$
$$P(0,1,i) = \frac{P(0,i)}{P(0,1)}d(i-1).$$

There are three yield curves two periods from now and each of them can be derived from the two yield curves one period from now:

$$P(2,2,i) = \frac{P(1,1,i)}{P(1,1,2)}u(i-2)$$
$$P(1,2,i) = \frac{P(0,1,i)}{P(0,1,2)}u(i-2) = \frac{P(1,1,i)}{P(1,1,2)}d(i-2).$$
$$P(0,2,i) = \frac{P(0,1,i)}{P(0,1,2)}d(i-2).$$

Although recursive substitutions give the above prices at time 2 directly computed from prices at time 0, from computational point of view this is not necessary, especially when pricing American contracts.

Using an arbitrage argument and solving a difference equation, Ho and Lee find the closed form solutions for u's and d's as follows:

$$u(k) = \frac{1}{\hat{p} + (1-\hat{p})\delta^k}$$

$$d(k) = \frac{\delta^k}{\hat{p} + (1-\hat{p})\delta^k}$$

where δ is a constant measuring the magnitude of the interest rate volatility. The higher the δ the higher the volatility. $\hat{p}$, as usual, represents the risk-neutral probability. With these two parameters, the whole binomial evolution can be identified. Then all contracts can be priced.

As an example, we assume δ to be 0.9 and $\hat{p}$ to be 0.6. From these two parameters, we can calculate ups and downs for various maturities as:

Time to Mat. (k)	$u(k)$	$d(k)$
0 year	1	1
1 year	0.9375	1.0417
2 years	0.8766	1.0823
3 years	0.8176	1.1216
4 years	0.7607	1.1595

We also assume an initial yield curve as:

Time to Mat.	Yield to Mat.
1 year	5%
2 years	6%
3 years	6.5%
4 years	6.8%

These yields can be easily transformed into bond prices:

Time to Mat.	$P(t,T)$	Bond Price
1year	$P(0,1)$	0.9524
2 years	$P(0,2)$	0.8900
3 years	$P(0,3)$	0.8278
4 years	$P(0,4)$	0.7686

From these bond prices, we can calculate forward prices: $\Psi(0,1,2)$, $\Psi(0,1,3)$, and $\Psi(0,1,4)$. Applying the corresponding up functions (with the matched, i.e., $u(1)$ for $\Psi(0,1,2)$, $u(2)$ for $\Psi(0,1,3)$, and $u(3)$ for $\Psi(0,1,4)$) to these forward prices, we can obtain a series of bond prices for the next period. Applying the corresponding down functions, we can obtain another series of bond prices. These two series of bond prices form a binomial uncertainty for the term structure.

By labeling the down state 0 and up state 1, we can create the two series of bond prices for the next period:

$P(0,1,i)$	Bond Price
$P(0,1,1)$	1.0000
$P(0,1,2)$	0.8761
$P(0,1,3)$	0.7619
$P(0,1,4)$	0.6598

$P(1,1,i)$	Bond Price
$P(0,1,1)$	1.0000
$P(1,1,2)$	0.9734
$P(1,1,3)$	0.9407
$P(1,1,4)$	0.9051

From these two term structures (up and down), we can derive the next period term structures and the number of the term structures will become 3. We again need to calculate forward prices at time 1: $\Psi(1,2,3)$ and $\Psi(1,2,4)$. These prices are calculated for both up state and down state. Again, applying corresponding up functions to these forward prices (i.e., $u(1)$ for $\Psi(1,2,3)$ and $u(2)$ for $\Psi(1,2,4)$) at both states to obtain two up term structures. Similarly, the down functions generate two down term structures. Note that the up term structure from the down state is identical to the up term structure from the down state – the basic binomial model. As a result, the number of states in time 2 is 3. Carrying out these calculations, we get:

$P(0,2,i)$	Bond Price	$P(1,2,i)$	Bond Price	$P(2,2,i)$	Bond Price
$P(0,2,2)$	1.0000	$P(1,2,2)$	1.0000	$P(2,2,2)$	1.0000
$P(0,2,3)$	0.8154	$P(1,2,3)$	0.9060	$P(2,2,3)$	1.0067
$P(0,2,4)$	0.6603	$P(1,2,4)$	0.8151	$P(2,2,4)$	1.0063

There is only forward price for each state at time 2: $\Psi(2,3,4)$. Applying ups and downs will yield 4 term structures for time 3 (each term structure has only one bond; the other bond expires):

$P(0,3,i)$	$P(1,3,i)$	$P(2,3,i)$	$P(3,3,i)$
$P(0,3,3) = 1.0000$	$P(1,3,3) = 1.0000$	$P(2,3,3) = 1.0000$	$P(3,3,3) = 1.0000$
$P(0,3,4) = 0.7592$	$P(1,3,4) = 0.8435$	$P(2,3,4) = 0.9372$	$P(3,3,4) = 1.0414$

Since the yield curve is fitted perfectly (or taken as an input), there is no pricing error in the Ho–Lee model. All models in this section (Section 2.4) share this assumption. The pricing is therefore for other contingent claims, for example, options. We can price a call option on a pure discount bond. For instance, we can use this bonimial model to find out the price of a call option that matures at time 2 and is written on a bond that matures at time 4 with a strike price of 0.80. Or we can calculate the futures price that matures at time 2 of a bond that matures at time 4.

The Ho–Lee model later on is understood as the first time-dependent parameter term structure model and it is shown that the Ho–Lee model has a closed form solution like Vasicek to the discount bond price. This may sound a little strange because in the original Ho–Lee model, bond prices are taken as given. If the price of a bond is taken as an input, then how can it be priced by a formula? The answer to this question is precisely the vehicle to unveil the black box of all time-dependent parameter models including Hull and White (1990) and Heath, Jarrow, and Morton (1992).

To see that the Ho–Lee model is actually a time-dependent parameter model, we first use the definition relationship between the forward price and the forward rate for one period:

$$f(i,n,n+1) = -\ln\frac{P(j,i,n+1)}{P(j,i,n)}$$

where j represents the state of economy at time i when the forward rate is observed. Substituting the previous bond prices from the previous down state, j–1, into both numerator and denominator, we get:

$$\begin{aligned}
-\ln\frac{P(j,i,n+1)}{P(j,i,n)} &= -\ln\frac{u(n+1-i)P(j-1,i-1,n+1)\,/\,P(j-1,i-1,i)}{u(n-i)P(j-1,i-1,n)\,/\,P(j-1,i-1,i)}\\
&= -\ln\frac{u(n+1-i)P(j-1,i-1,n+1)}{u(n-i)P(j-1,i-1,n)}\\
&= -\ln\frac{h(n+1-i)}{u(n-i)} + f(i-1,n,n+1)\\
&= f(i-1,n,n+1) + \ln\frac{\hat{p}+(1-\hat{p})\delta^{n+1-i}}{\hat{p}+(1-\hat{p})\delta^{n-i}}.
\end{aligned}$$

Also from the previous up state, we get:

$$\begin{aligned}
-\ln\frac{P(j,i,n+1)}{P(j,i,n)} &= -\ln\frac{d(n+1-i)}{d(n-i)} + f(i-1,n,n+1)\\
&= f(i-1,n,n+1) + \ln\frac{\hat{p}+(1-\hat{p})\delta^{n+1-i}}{\hat{p}+(1-\hat{p})\delta^{n-i}} - \ln\delta.
\end{aligned}$$

Combine the two equations to get:

$$f(i,n,n+1) = f(i-1,n,n+1) + \ln \frac{\hat{p} + (1-\hat{p})\delta^{n+1-i}}{\hat{p} + (1-\hat{p})\delta^{n-i}} - (1-\hat{p}) \ln \delta + \varepsilon_i$$

where the error term is:

$$\varepsilon_i = \begin{cases} (1-\hat{p}) \ln \delta & \text{if upstate at time } i \\ -\hat{p} \ln \delta & \text{if donwstate at time } i \end{cases}$$

which has an expected value of 0, $\hat{E}[\varepsilon] = 0$.

Also, moving the forward rate at time i–1 to the left hand side and letting the time interval between i and i–1 very small, we shall obtain a forward rate differential equation:

$$df = \alpha_f dt + \sigma_f d\hat{W}$$

where α_f and σ_f are constants. As to be demonstrated later, constant volatility of the forward rate process implies constant volatility of the spot rate while the drift of the spot rate remains arbitrary. In order to fit the yield curve, we need to make the drift term a time dependent function. Therefore, we can write the spot rate process for the Ho–Lee model as:

$$dr = \theta(t)dt + \sigma d\hat{W}.$$

No mean reversion

To be shown in the last chapter, this interest rate process is a stochastic *linear* differential equation and its solution is given in most mathematical textbooks (for example, Arnold (1974)): SDE: Theory & Applications, John Wiley

$$r(s) = r(t) + \int_t^s \theta(u)du + \int_t^s \sigma dW(u).$$

Assuming a similar risk premium to the Vasicek, we can find the price of the discount bond using the same risk-neutral expectation:

The Extended Vasicek model remains normally distribu

$$
\begin{aligned}
P(t,T) &= \hat{E}_t\left[\exp\left(-\int_t^T r(u)du\right)\right] \\
&= \exp\left(-r(t) - \int_t^T \int_t^s \hat{\theta}(u)duds + \frac{\sigma^2(T-t)^3}{6}\right) \\
&= D(t,T)e^{-r(t)(T-t)+\frac{\sigma^2(T-t)^3}{6}}.
\end{aligned}
$$

Note that the function D depends upon the time-dependent parameter θ. In order to fit perfectly the yield (price) curve, the time-dependent parameter θ needs to generate D for every point on the yield curve. In other words, θ is cooked in such a way that the discount bond pricing formula above is flexible enough to fit every point. The functional form of θ needs not be known.

It is interesting to examine the option formula. Note that since the short rate is normally distributed, the option formula remains the same. The only change is the bond return volatility:

$$
\text{var}[\ln P(T,s)] = \text{var}[r(T)] = \sigma^2(T-t)(s-T)^2.
$$

This option formula should yield the same option price as the original Ho–Lee model.

2.4.2 Hull–White Model

Realizing that time-dependent parameters can provide complete flexibility for curve fitting, Hull and White extend the Vasicek model in the following way. Take the extended Vasicek model for example:

$$
dr = \alpha(t)[\mu(t) - r]dt + \sigma dW.
$$

vs. $dr = \alpha(\mu - r)dt + \sigma dW$ Vasicek p. 26

Note that in this equation , σ is still constant. Although σ can be also time dependent, it will complicate the lattice model which calculates American prices. It will be shown in the last chapter that it is more convenient to price European contracts if σ is time dependent but not α.

It should be noted that parameters being time dependent does not change any distribution assumption of the process. The extended Vasicek model remains normally distributed. By Ingersoll (1989), we know that the market price of risk can be time dependent if parameters are time dependent. Therefore the previously defined q is now $q(t)$.

p 27

Hull and White obtain their solution to the bond price by solving a partial differential equation. Here, we shall take an approach that is consistent within our framework. Define a risk-neutral process as follows:

$$dr = [\alpha(t)\mu(t) - \sigma q(t) - \alpha(t)r]dt + \sigma d\hat{W}.$$

By Arnold, this stochastic linear equation has the following solution:

$$r(s) = \phi(s)\left(r(t) + \int_t^s \phi(u)^{-1}[\alpha(u)\mu(u) - \sigma q(u)]du + \int_t^s \phi(u)^{-1}\sigma d\hat{W}_u\right)$$

where $\phi(s) = \exp\left(-\int_t^s \alpha(u)du\right)$.

The solution to the bond price is quite complex, although it takes precisely the same form as the Vasicek original model. Recall the mean and variance of the state variable are:

$$\hat{E}_t[r(s)] = \phi(s)\left(r(t) + \int_t^s \phi^{-1}(u)\hat{q}(u)du\right)$$

and

$$\text{cov}_t[r(s), r(u)] = \phi(s)\left[\int_t^{\min(u,s)} \left(\phi^{-1}(w)\sigma\right)^2 dw\right]\phi(u)$$

and the term structure model will look like:

$$\begin{aligned} P(t,T) &= \hat{E}_t\left[\exp\left(-\int_t^T r(s)ds\right)\right] \\ &= e^{-m(t,T) + \frac{V(t,T)}{2}} \end{aligned}$$

where

$$\begin{aligned} m(t,T) &= \int_t^T \hat{E}[r(s)]ds \\ &= \int_t^T\left[\phi(s)\left(r(t) + \int_t^s \phi(u)^{-1}[\alpha(u)\mu(u) - \sigma q(u)]du\right)\right]ds \end{aligned}$$

and

$$\begin{aligned} V(t,T) &= \int_t^T \int_t^s 2K_t[r(s), r(u)]duds \\ &= \int_t^T \int_t^s 2\phi(s)\left[\int_t^u \left(\phi^{-1}(w)\sigma\right)^2 dw\right]\phi(u)duds. \end{aligned}$$

It is seen that unless the functions $\alpha(t)$ and $\mu(t)$ are in closed form, we cannot arrive at the closed form for the bond price.

Note in this model, there are two pieces of flexibility, one in $\alpha(t)$and the other in $\mu(t)$. This means that the pricing formula can generate any bond prices to match the observed ones traded in the marketplace. In fact, one piece of flexibility is enough to generate all wanted bond prices. The other flexibility is used to fit option prices that reflect volatilities in the future. We shall now look at the option formula of this model. Since the bond price, no matter how complex it is, remains a log normal distribution. As a result, its option pricing formula is:

$$C(t, T_c, T) = P(t, T)N(d) - P(t, T_c)KN(d - \sqrt{V_p})$$

where

$$d = \frac{\ln(P(t,T)/KP(t,T_c) + V_p/2}{\sqrt{V_p}}$$

and

$$\begin{aligned} V_p &= \text{var}(\ln P(T_c, T)) \\ &= F(T_c, T)^2 \,\text{var}(r(T_c)) \\ &= F(T_c, T)^2 \phi(T_c)^2 \sigma^2 \textstyle\int_t^{T_c} \phi(w)^{-2} dw. \end{aligned}$$

Note that this formula does not depend on $\mu(t)$ and only depends on $\phi(t)$ and σ. If σ is given, one can use the flexibility provided by $\alpha(t)$ to fit the option prices and then jointly with the other flexibility provided by $\mu(t)$ to fit the yield curve.

Although the bond and option formulas are presented in an analytical format, since the model has to fit all the yields (of bond prices) and all the volatilities (of option prices), the model can only be used if there is a numerical model, say lattice model. Integrals of unknown functions make the model difficult to use.

If the model is not required to fit the volatility curve, then the solution becomes so much easier. We need only one time-dependent parameter. From the experience above, we know that the parameter which needs to be time dependent is μ. Although time-dependent α will serve the same purpose, it unnecessarily complicates the option formula. We rewrite the bond and option formulas as follows:

$$\begin{aligned} P(t, T) &= \hat{E}_t\left[\exp\left(-\textstyle\int_t^T r(s)ds\right)\right] \\ &= e^{-m(t,T) + \frac{V(t,T)}{2}} \end{aligned}$$

where

$$\begin{aligned} m(t, T) &= r(t)F(t, T) + \textstyle\int_t^T \left[e^{-\alpha(s-t)}\left(\int_t^s e^{\alpha(u-t)}[\alpha\mu(u) - \sigma q(u)]du\right)\right]ds \\ &= r(t)F(t, T) + D(t, T) \end{aligned}$$

and

$$V(t, T) = \frac{\sigma^2}{\alpha^2}\left[T - t - \frac{e^{-2\alpha(T-t)}}{2\alpha} + 2\frac{e^{\alpha(T-t)}}{\alpha} - \frac{3}{2\alpha}\right].$$

It is quite clear that the function D is used to provide the flexibility to fit all necessary bond prices. Since the time-dependent functions μ and q do not enter into the option formula, the fitting of the yield curve can be done through changing D with no impact on the option prices.

2.4.3 Black–Derman–Toy Model

Black, Derman, and Toy (1990), abbreviated as BDT, use the same logic as Ho and Lee and develop a binomial model that uses traded prices as given. Their model differs from the Ho–Lee model in that their model in addition to the yield curve also fits the volatility curve. The distribution assumption of the short rate is log normally distributed. Immediately after the discrete model, Black and Karazinski (1991) develop a continuous time version of the BDT model The Black–Karazinski model is no different from the Hull–White given above in the sense that it is a numerical model that fits both the yield and the volatility curves. It differs, however, in the distribution assumption. It is log normal and the extended Vasicek is normal.

The BDT model assumes a binomial evoluation of the short rate in the same manner as the binomial model for the stock option by Cox, Ross, and Rubinstein (1979). The risk-neutral probabilities for up and down is fixed at 1/2. They adjust the up and down jumps to assure the perfect matches on the bond prices and volatilities. To illustrate, we define a yield curve and volatility curve as:

Maturities	Yields	Yield Volatilities
1	10%	20%
2	11	19
3	12	18
4	12.5	17
5	13	16

The one-year bond price is certain because the one-year interest rate is known. A discrete discounting with 10% is 0.9091. The two-year bond will have two possible bond prices in a year:

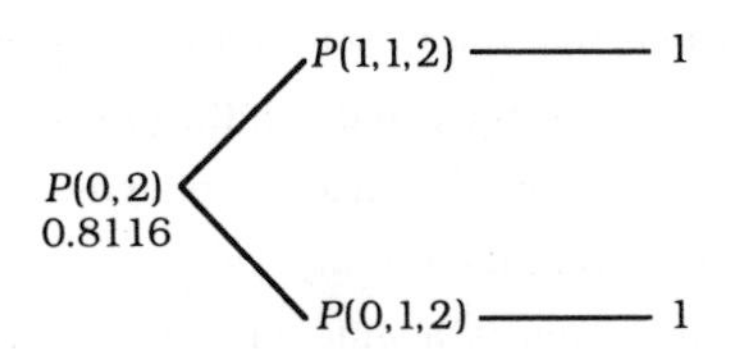

The two-year bond price which is discounted discretely at 11% is 0.8116. We know that this price also equals:

$$P(0,2) = \frac{P(0,1,2) + P(1,1,2)}{2} \times P(0,1)$$
$$0.8116 = \frac{1}{2}\left(\frac{1}{1+r_{10}} + \frac{1}{1+r_{11}}\right) \times 0.9091.$$

To solve for the two interest rates (up and down), we need another equation. This is where the volatility measure comes in. In the Cox–Ross–Rubinstein binomial model, we understand that the up and down are characterized by:

$$\begin{cases} u = e^{\sigma\sqrt{\Delta t}} \\ d = e^{-\sigma\sqrt{\Delta t}} \end{cases}.$$

Solving for σ, we obtain:

$$\sigma = \frac{\ln \frac{u}{d}}{2} = \frac{\ln \frac{r_0 u}{r_0 d}}{2} = \frac{\ln \frac{r_{11}}{r_{10}}}{2} = 19\%.$$

Solving these simultaneous equations, we can get:

$$\begin{cases} r_{11} = 14.32\% \\ r_{10} = 9.79\% \end{cases} \text{ or } \begin{cases} u = 1.209 \\ d = 0.827 \end{cases}.$$

This implies the two bond prices are:

$$P(1,1,2) = \frac{1}{1+r_{11}} = \frac{1}{1+14.32\%} = 0.8747$$
$$P(0,1,2) = \frac{1}{1+r_{10}} = \frac{1}{1+9.79\%} = 0.9108.$$

One thing interesting to note is that the r_0 solved from these equations is not equal to 10%. It is 11.84%. In other words, the u and d in the problem are not up and down for the initial interest rate. They are obtained by matching the yield and the yield volatility. What this implies is that the two interest rates at time 1 are not conditional on the initial interest rate. This is a different notion than a stochastic process commonly understood.

For the three year bond, again we first find its price using the three year yield to be 0.7118. Therefore,

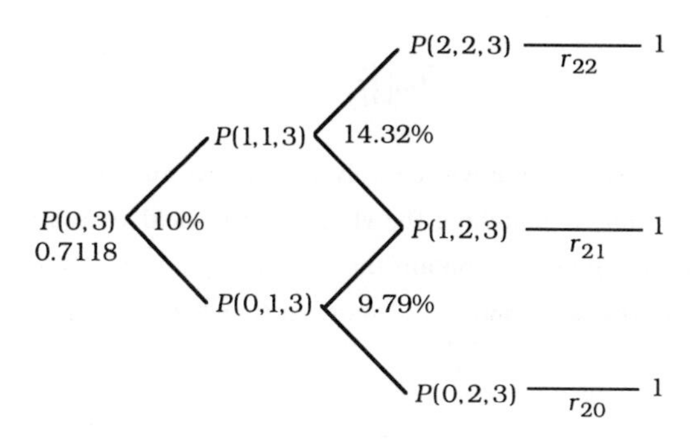

Similarly, we know that this bond price can be obtained by discounting the payoff:

$$\begin{aligned}
P(0,3) &= \frac{\frac{1}{2}P(1,1,3)+\frac{1}{2}P(0,1,3)}{1+10\%} \\
&= \frac{1}{1.1}\times\frac{1}{2}\times\left[\frac{\frac{1}{2}P(2,2,3)+\frac{1}{2}P(1,2,3)}{1.1432}+\frac{\frac{1}{2}P(1,2,3)+\frac{1}{2}P(0,2,3)}{1.0979}\right] \\
&= \frac{1}{1.1}\times\frac{1}{4}\times\left[\frac{\frac{1}{1+r_{22}}+\frac{1}{1+r_{21}}}{1.1432}+\frac{\frac{1}{1+r_{21}}+\frac{1}{1+r_{20}}}{1.0979}\right].
\end{aligned}$$

The second volatility is matched by the yield volatility of the two year bond at time 1:

$$18\% = \frac{\ln\frac{\sqrt{1/P(1,1,3)}-1}{\sqrt{1/P(0,1,3)}-1}}{2}$$

This equation should again be expressed in terms of r_{22}, r_{21}, and r_{20}. We have three unknowns to solve for and so far we have only two equations. A third equation is imposed to ensure that u and d are only time dependent but not state dependent. In other words, the product of r_{22} and r_{20} should equal r_{21}^2. With three equations, we can then solve for the three interest rates. It is clear that solving for these three interest rates is not easy, since all three equations are not linear. It should also be aware that at time 3, there will be a system of 4 simultaneous equations to solve. As the number of periods grows, the solving of the system can be costly.

There is an easy way to avoid solving a growing system. For any period, since u

and d need to be the same across states, we can always express any interest rate as a function of the bottom rate:

$$r_{ij} = r_{i0}\left(\frac{u}{d}\right)^{j} = r_{i0}u^{2j}$$

Therefore, for any period, there will be only 2 variables to solve for. And these two variables are solved for by matching the yield and the yield volatility. Solving for the three interest rates at time 2, we obtain the following two period tree. Note again that the u and d that are the same across the states at time 2 are not applied to 14.32% and 9.79%.

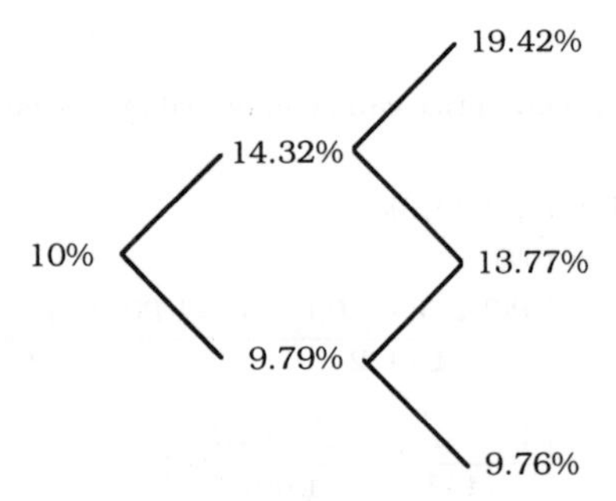

Repeating this procedure, we shall calculate all rates by matching yields and volatilities. The details and results can be found in BDT's original work. Note that this binomial model is exactly identical to the Cox–Ross–Rubinstein except that the u's and d's are all different for different periods except that future interest rates do not depend upon previous interest rates.

It is clear that the distribution of interest rates in each period is log normal. However, the continuous time limit is not clear and a diffusion process for the interest rate cannot be easily defined. A continuous time mean reverting log normal model is given by Black and Karazinski (1991). Hull and White argue that this model can be modeled with a trinomial model which is more efficient.

2.4.4 Heath–Jarrow–Motorn Model

The Heath, Jarrow, and Morton (1992) model, abbreviated as HJM, to some extent, is a generalization of the continuous time version of the Ho–Lee model. They instead of the short rate, model forward rates which is the rates of return of the forward prices. However, the HJM model is parametized by time-dependent parameters and incorporates mean reversion. Although the model looks different from all previous models, its implementation remains exactly the same as all the other time dependent models. We shall be more specific later but let us look at the model itself.

The HJM model is, strictly speaking, not a pricing model, but a framework under which all arbitrage-free term structure models can be derived. However, HJM do provide two explicit models but both of which are normally distributed. Solutions to non-normal assumptions need be done through numerical methods.

HJM instead of modeling the short rate, model the forward rate:[8]

$$df(t,T) = \alpha_f(t,T)dt + \sigma_f(t,T)d\hat{W}.$$

The solution of this differential is simple:

$$f(t,T) = f(0,T) + \int_0^t \alpha(u,T)du + \int_0^t \sigma(u,T)d\hat{W}(u).$$

The short rate which is equal to $f(t,t)$ can be:

$$r(t) = f(0,t) + \int_0^t \alpha(u,t)du + \int_0^t \sigma(u,t)d\hat{W}(u).$$

If we look at the bond price dynamics, by Ito's lemma on the short rate, we know that:

$$dP(t,T) = P(t,T)r(t)dt + P(t,T)\sigma_P d\hat{W}(t)$$

where

$$\sigma_P = \frac{\partial P}{\partial r}\frac{\sigma_r}{P}.$$

By Ito's lemma, we can write the dynamics of the log of the bond price:

$$d\ln P(t,T) = \left(r - \tfrac{1}{2}\sigma_P^2\right)dt + \sigma_P d\hat{W}(t).$$

Now, the forward rate is defined in terms of the bond price as follows:

[8] We use a single factor version of the HJM model.

$$f(t,T) = -\frac{d\ln P(t,T)}{dT}$$
$$= -\frac{\ln P(t,T+dT) - \ln P(t,T)}{dT}.$$

Therefore,

$$df(t,T) = -\frac{d\ln P(t,T+dT) - d\ln P(t,T)}{dT}$$
$$= \frac{\sigma_P^2(t,T+dT) - \sigma_P^2(t,T)}{2(dT)}dt + \frac{\sigma_P(t,T+dT) - \sigma_P(t,T)}{dT}dW(t)$$
$$= \sigma_P(t,T)\frac{d\sigma_P(t,T)}{dT}dt + \frac{d\sigma_P(t,T)}{dT}dW(t).$$

It is clear that:

$$\sigma_f(t,T) = \frac{d\sigma_P(t,T)}{dT}$$
$$\alpha_f(t,T) = \sigma_P(t,T)\frac{d\sigma_P(t,T)}{dT} = \sigma_f(t,T)\sigma_P(t,T)$$

which means the forward rate drift cannot be arbitrary and needs to be a function of the forward rate volatility and bond return volatility. Substituting in the explicit form of $\sigma_P(t,T)$:

$$\sigma_f(t,T) = \frac{d[\partial P(t,T)/\partial r]}{dT}\frac{\sigma_r(t,T)}{P(t,T)}$$
$$\alpha_f(t,T) = \left[\frac{d[\partial P(t,T)/\partial r]}{dT}\right][\partial P(t,T)/\partial r]\left[\frac{\sigma_r(t,T)}{P(t,T)}\right]^2.$$

This result also indicates that since the drift term does not affect the forward rate process, it cannot affect any claim prices. It also gives the relation between the short rate parameters and forward rate parameters. From the expression above, we know that since the bond price is a function of time to maturity, the forward rate volatility needs also to be a function of time to maturity. Suppose we let the forward rate volatility be $\sigma_f(t,T) = b_f\sqrt{T-t}$ and also the short rate volatility be $\sigma_r(t,T) = b_r\sqrt{T-t}$. Then, we obtain:

$$
\begin{aligned}
dP_r / dT &= b_r / b_f = b \\
P_r &= b(T-t) \\
dP / dr &= b(T-t) \\
P(t,T) &= b(T-t)r + 1
\end{aligned}
$$

The bond price solution is shown to be a lilnear function of the short rate! Certainly this example is inappropriate because it contradicts the deterministic limit. A more realistic example would be to set the forward rate volatility as $\sigma_f(t,T) = b_f P$ and the spot rate volatility (for simplicity) as a constant, b_r. Then we get:

$$
\begin{aligned}
dP_r / dT &= \sigma_f / \sigma_r = bP \\
P_r / P &= b(T-t) \\
d\ln P &= b(T-t)dr \\
P(t,T) &= \exp(b(T-t)r)
\end{aligned}
$$

Now, this example will converge to the deterministic limit. These two examples demonstrate that the forward rate volatility cannot be arbitrarily determined, or conflicts may take place.

2.4.5 Relations Among Time-Dependent Parameter Models

Other than different distribution assumptions: Hull and White use normal, HJM use normal, BDT use log normal, all time dependent models adjust the parameters to fit the yield curve and the volatility curve. Although the HJM model parametizes the forward rates, the model can be implemented with the spot rate lattice model.

The easiest way to see that these models are all similar to one another is to construct an example. Let us start from the Hull–White model:

$$dr = (a(t) - \alpha(t)r)dt + \sigma_r d\hat{W}.$$

Therefore the bond price needs to satisfy the following equation:

$$dP(t,T) = r(t)P(t,T)dt + \tfrac{P_r(t,T)\sigma_r}{P(t,T)} P(t,T)d\hat{W}(t).$$

We know from above that:

$$P_r(t,T) = -P(t,T)\int_t^T \phi(s)ds.$$

And we then know that:

$$\sigma_f(t,T) = \sigma_r \frac{d\int_t^T \phi(s)ds}{dT} = \sigma_r \phi(T) = \sigma_r \exp\left(\int_t^T -\alpha(s)ds\right).$$

It is then quite simple to translate from one model to the other. The similarity between the BDT model and the HJM model can be analyzed in the same manner.

It has been seen by now that all time-dependent parameter models are similar in the sense that they jiggle the parameters enough to fit the observed yield and volatility curves. Under the normality assumption, all models can be made completely equal to one another if parameters are set up in a certain way. Since all models are the same, in pricing American contracts where numerical methods become necessary, we need to choose the simplest way to build the model. Then, it becomes clear that the Hull White model is the most suitable for the lattice model. The Hull–White model relies on only one single variable, the short rate while the HJM or the BDT model requires a complete set of points on the forward rate curve or the spot rate curve.

2.5. COUPON BOND

Discount bonds are present values of \$1. The present value of any payoff is therefore the payoff multiplied by the discount bond price. If this logic applies to coupon bonds, the coupon bond price is therefore the sum of discounted coupons (the last cash flow includes the face value). Formally,

$$Q(t,\underline{T}) = \Sigma_{i=1}^n c_i P(t,T_i) + 100P(t,T_n)$$

where $\underline{T}$ is a vector of all coupon arrival times.

Futures price is a risk-neutral expectation of its payoff at maturity.

CHAPTER 3
OPTIONS AND FUTURES

3.1 FUTURES AND FORWARD

A bond futurescontract gives both the long and the short a guaranteed price to buy and sell the bond. Therefore, determining a futures price at the time both parties sign the contract is a pricing problem. However, the futures price is not a price about the futures contract. Readers should keep in mind that the futures contract itself should have no value, since it is merely a promise to exchange money for goods. Because the futures price is not a price of an asset, the "pricing" is slightly different of the bond shown previously. Aside from complex mathematical or arbitrage treatments, as we understand it now, the futures price is a risk neutral expectation of its payoff at maturity. This differs from other valuation problems in that there is no cash changing hands when trading futures contracts.

Suppose a futures contract matures at time T_f upon which a bond that matures at T will be delivered. Then the futures price to be written on today's contract is:

$$\Phi(t, T_f, T) = \hat{E}_t[P(T_f, T)].$$

Under the Vasicek model, the distribution of $r(T_f)$ is normal with mean and variance given earlier. Therefore, using the moment generating function of a normal variate and rearranging terms give a futures price as follows:

$$\begin{aligned}\Phi_{vas}(t, T_f, T) &= \hat{E}\left[e^{-r(T_f)F(T_f,T)-G(T_f,T)}\right] \\ &= e^{-G(T_f,T)}e^{-F(T_f,T)\hat{E}[r(T_f)]+\frac{F(T_f,T)^2 \hat{V}[r(T_f)]}{2}} \\ &= e^{-r(t)X(t,T_f,T)-Y(t,T_f,T)}\end{aligned}$$

where

$$X(t, T_f, T) = F(t, T) - F(t, T_f)$$

$$Y(t, T_f, T) = \left(\mu - \frac{q\sigma}{\alpha} - \frac{\sigma^2}{2\alpha^2}\right)[T - T_f - X(t, T_f, T)] - \frac{\sigma^2}{2\alpha^2}\left[X(t, T_f, T) - \frac{\alpha}{2}X^2(t, T_f, T) - F(T_f, T)\right]$$

The CIR formula of the square root process is more complex. However, unlike the bond price solution that requires solving the Laplace transform, the solution to the futures price needs the moment generating function of a transformed non-central chi-square variate. The futures price is:

$$\Phi_{cir}(t,T_f,T) = \hat{E}\left[A(T_f,T)e^{-r(T_f)B(T_f,T)}\right]$$
$$= \hat{E}\left[A(T_f,T)e^{-x(T_f)B(T_f,T)/(2\hat{c})}\right]$$
$$= A(T_f,T)\frac{\exp\left(\frac{-B(T_f,T)\hat{\lambda}/(2\hat{c})}{1+B(T_f,T)/\hat{c}}\right)}{(1+B(T_f,T)/\hat{c})^{4\kappa\theta/v^2}}$$

where definitions of the notation are given previously. The second line is to transform r into a standard non-central chi-square variate and from the moment generating function of the third line, we arrive at the final solution:

$$\Phi_{cir}(t,T_f,T) = C(t,T_f,T)e^{-rD(t,T_f,T)}$$

where

$$C(t,T_f,T) = \left(\frac{\hat{c}}{\hat{c}+B(T_f,T)}\right)^{2\kappa\theta/v^2} A(T_f,T)$$

and

$$D(t,T_f,T) = \frac{\hat{c}e^{-(\kappa+\lambda)(T_f-t)}}{\hat{c}+B(T_f,T)}B(T_f,T).$$

It is noted that the solution remains the same exponential form as the bond under either assumption. Therefore, like bonds, as long as the parameters are known, the prices for all maturities can be readily calculated.

[EXAMPLE 1]

Taking parameter values from the previous chapter and further assuming that the futures expiration date is in a year and the underlying discount bond is a 5-year bond, we can compute the futures prices under both Vasicek and CIR models. The substitutions are straightforward and the results are $73.55 for the Vasicek model and $73.33 for the CIR model. It is seen that the difference between the futures prices are not large.

Though two models look similar and are derived similarly, the Vasicek version of the futures price can be expressed as preference-free, i.e., the risk parameter q will drop out of the equation. To see that, we shall re-express the Vasicek version of the futures price as follows:

$$\Phi(t, T_f, T) = \frac{P(t,T)}{P((t,T_f)} e^J$$

where

$$J = -\frac{1}{2}\sigma^2 F(t, T_f)^2 F(T_f, T).$$

Note that J does not contain q. Therefore, the above equation is preference-free. The risk parameter is hidden in bond prices which are now taken from the market and not computed by the pricing formula. This result is analogous to the Black–Scholes option formula where the drift drops out of the equation.

The forward price is usually found by a pure arbitrage. This is because its payoff can be completely duplicated. As mentioned in Chapter 1, this qualifies a pure arbitrage in which the risk is completely diminished. Since the forward contract has a linear payoff at maturity, it is equivalent to a long call and a short put with the forward price as their strike price. Since the forward contract has no value, the call premium will have to equal the put premium. As a consequence, the forward price is the strike price that will make two premiums the same. Formally,

$$O_c(t, T_f, T) - O_p(t, T_f, T) = 0.$$

By the put–call parity, this is equivalent to:

$$P(t,T) - P(t, T_f)\,K = 0$$

$$\frac{P(t,T)}{P(t,T_f)} = K = \Psi(t, T_f, T).$$

It is seen that the forward formula is independent of the number of factors since it is model-free. Any contract that can be priced by pure arbitrage would have this property. Market prices can price among themselves.

Given that the forward price is a ratio of two different-maturity bonds, we can then write the Vasicek futures price as[9]:

$$\Phi(t, T_f, T) = \Psi(t, T_f, T)e^J.$$

We know that forward and futures prices should be equal when there is no marking to market effect or when the interest rate is deterministic. We see that when σ=0, interest

[9]However, there is no easy relationship between the futures and forward prices under the CIR model.

rates become deterministic, then J becomes 0 and the two prices are equal.

Despite solutions to forward and futures under both models are simple to use, they are not commonly seen traded. What are commonly traded are the coupon bond futures like Treasury bond futures. To value a coupon bond futures, we first write the coupon bond as a portfolio of discount bonds, or alternatively, as a sum of discounted coupons:

$$Q(t, \underline{s}) = \Sigma_{j=1}^{n} c \cdot P(t, T_j) + 100 \cdot P(t, T_n)$$

where c is the per \$100 coupon payment. We use $\underline{s}$ as the collection of all maturities from 1 to n.

The futures price is simple. Recall the risk neutral expectation we have used. The futures price is the risk neutral expectation of the future bond price. Since now the bond price is Q, we shall just take an expected value of Q in the future time:

$$\begin{aligned}\Phi(t, T_f, \underline{s}) &= \hat{E}_t\left[Q(T_f, \underline{s})\right] \\ &= \hat{E}_t\left[\Sigma_{j=1}^{n} c \cdot P(T_f, T_j) + 100 \cdot P(T_f, T_n)\right] \\ &= \Sigma_{j=1}^{n} c \cdot \hat{E}_t[P(T_f, T_j)] + 100 \cdot \hat{E}_t[P(T_f, T_n)] \\ &= \Sigma_{j=1}^{n} c \cdot \Phi(t, T_f, T_j) + 100 \cdot \Phi(t, T_f, T_n).\end{aligned}$$

For the multi-factor models, because the factors are independent, the discount bond solution can be worked out the same way as the bond, i.e., a product of two single-factor futures pricing formulas:

$$\begin{aligned}\Phi(t, T_f, T) &= \hat{E}[P_1(T_f, T) P_2(T_f, T)] \\ &= \hat{E}[P_1(T_f, T)]\hat{E}[P_2(T_f, T)] \\ &= \Phi_1(t, T_f, T)\Phi_2(t, T_f, T).\end{aligned}$$

3.2 BOND OPTIONS

The easiest and the most basic interest rate option is the option on pure discount bond. Since the pure discount bond does not pay coupons, like stock option with no dividends, the European value is also applicable to American options since the early exercises are worthless. This option also serves as a very basic valuation formula to other more complex formulas such as coupon bond options, options on yields, futures options, caps and floors, swaptions, and numerous other options. To derive its formula, we need a somewhat sophisticated mathematics called change of measure. If readers find the mathematical treatments difficult to grasp, they can skip the details because they have nothing to do with understanding the contract. We need them only when we need to price other exotic interest rate contracts.

At the maturity of the option, the contract will deliver a pure discount bond as an exchange for cash of the amount of the exercise price, if the option is in the money. Formally, the payoff can be written as:

$$O_c(T_c, T_c, T) = \max\{P(T_c, T) - K, 0\}$$

where K is the strike price. From above, we know that a European value of the call is merely a risk neutral expectation of this payoff. I.e.,

$$\begin{aligned} O_c(t, T_c, T) &= \hat{E}\left[\exp\left(-\int_t^{T_c} r(s)ds\right)\max\{P(T_c, T) - K,\ 0\}\right] \\ &= \hat{E}\left[\exp\left(-\int_t^{T_c} r(s)ds\right)(P(T_c, T) - K)\ 1_{\{P>K\}}\right] \\ &= \hat{E}\left[\exp\left(-\int_t^{T_c} r(s)ds\right)P(T_c, T)\ 1_{\{P>K\}}\right] - K\ \hat{E}\left[\exp\left(-\int_t^{T_c} r(s)ds\right)1_{\{P>K\}}\right] \end{aligned}$$

where $1_{\{\cdot\}}$ is an indicator function. What it does is a notational convenience to substitute for the max function. However, this indicator function has a nice property which is its expectation represents probability of $P > K$.

In the following we need to perform the "change of measure" twice. The purpose is to make the calculation easier and utilize the nice property of the indicator function. The change of measure enables us to separate the product inside the expectation. For example,

$$\begin{aligned} O_c(t, T_c, T) &= \hat{E}_t\left[\exp\left(-\int_t^{T_c} r(s)ds\right)P(T_c, T)\right]\tilde{E}_t\left[1_{\{P>K\}}\right] - K\ \hat{E}\left[\exp\left(-\int_t^{T_c} r(s)ds\right)\right]\overline{E}_t\left[1_{\{P>K\}}\right] \\ &= \hat{E}_t\left[\exp\left(-\int_t^{T_c} r(s)ds\right)P(T_c, T)\right]\tilde{\Pr} - K\hat{E}_t\left[\exp\left(-\int_t^{T_c} r(s)ds\right)\right]\overline{\Pr} \end{aligned}$$

To have a nice closed form solution like Black–Scholes, we need a simple assumption for the distribution for the bond price, P. So far it has been shown that only Vasicek and CIR can provide such simple solutions. While the expected value in the second term above is naturally the bond price, $P(t, T_c)$, we know from the Appendix that the first term above is:

$$
\begin{aligned}
&\hat{E}_t\left[\exp\left(-\int_t^{T_c} y(s)ds\right)P(T_c,T)\right]\tilde{\Pr} \\
&= \hat{E}_t\left[\exp\left(-\int_t^{T_c} y(s)ds\right)\right]\overline{E}_t[P(T_c,T)]\tilde{\Pr} \\
&= P(t,T_c)\frac{P(t,T)}{P(t,T_c)}\tilde{\Pr} \\
&= P(t,T)\tilde{\Pr}
\end{aligned}
$$

Therefore, the solution should look like:

$$O_c(t,T_c,T) = P(t,T)\tilde{\Pr} - KP(t,T_c)\overline{\Pr}.$$

Note that this is a general solution. We have not imposed any assumption on the interest rate, or the bond price. In the case of Vasicek, note that the future bond price at option's maturity is log normally distributed (because the future interest rate is normally distributed and the bond price is an exponential function of it), the single factor model for the option can follow exactly the Black–Scholes derivation and becomes:

$$O_c(t,T_c,T) = P(t,T)\,N(d) - P(t,T_c)\,K\,N(d-\sqrt{V_p})$$

where

$$d = \frac{1}{\sqrt{V_p}}\left(\ln\frac{P(t,T)}{P(t,T_c)\,K} + \frac{V_p}{2}\right)$$

$$V_p = \mathrm{var}[\ln P(T_c,T)] = F(T_c,T)^2\,\frac{\sigma^2\left(1-e^{-2\alpha(T_c-t)}\right)}{2\alpha}$$

Although this formula is given by Jamshidian in 1989, we call it the Vasicek option model in order to make a clearer reference to the CIR option model. It is interesting to note that this formula has been given simultaneously by three different authors, Chaplin, Jamshidian, and Rabinovitch. Chaplin (1987), although never published his work, solves the partial differential equation beautifully and obtains the solution. Since his partial differential equation depends on the interest rate, not the bond price, his solution, although exactly the same, looks a little different from the other two papers. Rabinovitch (1989) uses the Merton approach and obtains the same result as Jamshidian, only after a slight correction by Chen (1991). Jamshidian, of course, solves the problem the way we use in the book.

The form used above uses explicitly the property of log normality. The resemblance between this formula and the Black–Scholes formula is clear. This is because here in the Vasicek option model the bond price follows a log normal distribution, just like the stock price. The difference is that the volatility, V_p, is not constant. This is because bond prices will converge to the face value while stock prices will not. The convergence tends to reduce the volatility towards maturity. To see that, we use parameter values from the previous chapter:

α	0.2456
σ	0.0289
T	5 years

The volatility will look like the following graph for the next five years till the bond expires.

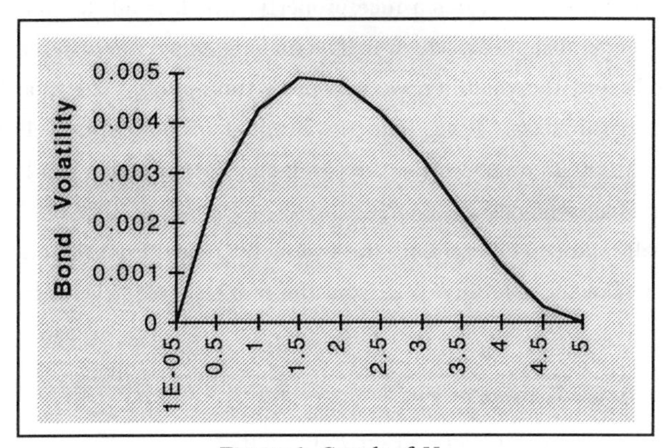

Figure 1. Graph of V_p

The volatility curve will first go up in that distant bond prices are less certain than near bond prices. However, since the bond price will converge to the face value with certainty, its volatility will go down. This is why the volatility curve will first go up and then go down. This is a sharp comparison to the Black–Scholes case where the stock return volatility is a positively sloped straight line. This is because in the Black–Scholes, the annualized volatility is a constant. To find the annualized volatility curve, we divide V_p by the number of years and obtain the following downward sloping curve for the volatility. We should note that the rate volatility is constant over time. This downward sloping curve is certainly driven by maturity.

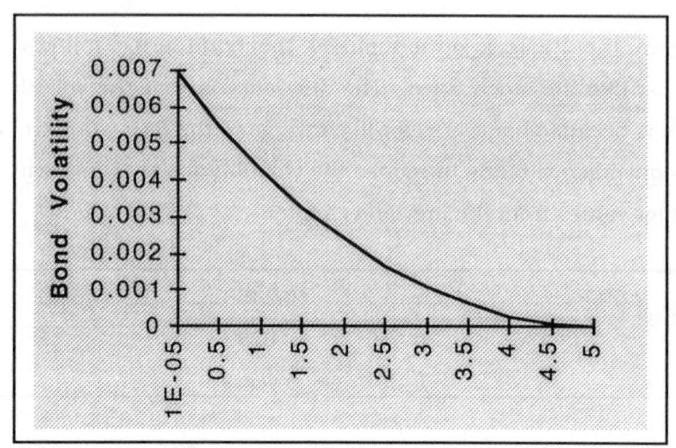

Figure 2. Graph of $V_p / (T_c - t)$

Vasicek's option model has a nice property, which is all the parameters in the term structure model disappear, which means as long as we can observe market prices of the two bonds and the volatility, we can price the option. The model is therefore independent of parameters. If so, then the number of factors as well as parameter values do not matter. In other words, the two factor option model will look exactly the same as this one factor model.

If we really want to differentiate the two factor model from the one factor model, then we cannot take the volatility, V_p, from the market. We have to calculate it from the formula:

$$\begin{aligned} V_p &= \text{var}[\ln P(T_c, T)] \\ &= \text{var}\left[-y_1 F_1(T_c, T) - y_2 F_2(T_c, T)\right] \\ &= F_1(T_c, T)^2 \frac{\sigma_1^2\left(1 - e^{-2\alpha_1(T_c - t)}\right)}{2\alpha_1} + F_2(T_c, T)^2 \frac{\sigma_2^2\left(1 - e^{-2\alpha_2(T_c - t)}\right)}{2\alpha_2} \end{aligned}$$

It is seen that if we calculate the volatility from the formula, then alpha's and sigma's will both affect the bond return volatility. Of course, if we substitute in the bond pricing formula of the two-factor model, then all parameters will come in. However, normally we would have to use real prices for the bonds. The reason is that if the two-factor term structure model cannot generate correct prices for the bonds, then the model is questionable. Finally, it should be noted that the Hull–White model with constant speed parameter gives exactly the formula as the Vasicek model.

The formula under the CIR framework can be derived similarly; only that the two probabilities are much more complex. Interested readers can follow the change of measure technique to derive the CIR option model. It should be noted that the bond price distribution is not known. As a result, the distribution functions are written in terms of the interest rate. This is just like the approach taken by Chaplin (1987). Here, we only presents the formula in its final form:

$$O_c(t,T_c,T) = P(t,T)\chi^2\left(2(\phi+\psi+B(T_c,T))r^*; \frac{4\kappa\theta}{\upsilon^2}, \frac{2\phi^2 e^{\gamma(T_c-t)}r^*}{\phi+\psi+B(T_c,T)}\right)$$
$$+KP(t,T_c)\chi^2\left(2(\phi+\psi)r^*; \frac{4\kappa\theta}{\upsilon^2}, \frac{2\phi^2 e^{\gamma(T_c-t)}r}{\phi+\psi}\right)$$

where

$$\phi = \frac{2\gamma}{\upsilon^2\left(e^{\gamma(T_c-t)}-1\right)}$$
$$\psi = (\kappa+\lambda+\gamma)/\sigma^2$$
$$r^* = \ln[A(T_c,T)/K]/B(T_c,T).$$

The formula for the two-factor CIR option prices is very complicated. This is because the linear combination of two non-central chi-square variates is not known. From our general option model,

$$O_c(t,T_c,T) = P(t,T)\tilde{\Pr} - KP(t,T_c)\overline{\Pr}$$

where

$$\tilde{\Pr} = \iint_{P>K} \tilde{\varphi}(y_1)\tilde{\varphi}(y_2)dy_1dy_2$$
$$\overline{\Pr} = \iint_{P>K} \overline{\varphi}(y_1)\overline{\varphi}(y_2)dy_1dy_2.$$

Note that in CIR's two factor model, it is impossible to reduce the double integrals into single integrals because the distribution of the linear combination of two non-central chi-square variables is unknown:

$$P > K$$
$$-y_1B_1(T_c,T) - y_2B_2(T_c,T) > -\overline{y}_1B_1(T_c,T) - \overline{y}_2B_2(T_c,T)$$

Chen and Scott (1992) have solved this problem and give a formula like:

$$O_c(t,T_c,T) = P(t,T)\chi^2(2L_1^*y_1^*, 2L_2^*y_2^*, \delta_1, \delta_2, \Lambda_1^*, \Lambda_2^*)$$
$$-KP(t,T_c)\chi^2(2L_1^\circ y_1^*, 2L_2^\circ y_2^*, \delta_1, \delta_2, \Lambda_1^\circ, \Lambda_2^\circ)$$

where

$$L_i^* = \phi_i + \psi_i + B_i(T,s)$$

$$L_i^{\circ} = \phi_i + \psi_i,$$
$$\delta_i = \frac{4\kappa_i\theta_i}{v_i^2}$$
$$\Lambda_i^* = \frac{2\phi_i^2 e^{\gamma_i(T-t)} y_i}{\phi_i + \psi_i + B_i(T,s)}$$
$$\Lambda_i^{\circ} = \frac{2\phi_i^2 e^{\gamma_i(T-t)} y_i}{\phi_i + \psi_i}$$
$$y_i^* = \frac{\ln A_1(T_c,T) + \ln A_2(T_c,T) - \ln K}{B_i(T_c,T)}.$$

Each chi-squared probability function is a bivariate non-central chi-squared probability defined as follows:

$$\chi^2(L_1, L_2, \delta_1, \delta_2, \Lambda_1, \Lambda_2) = \int_0^{L_2} F\left(L_1 - \frac{L_1}{L_2} x_2; \delta_1, \Lambda_1\right) f(x_2; \delta_2, \Lambda_2) dx_2$$

where $F(\cdot)$ and $f(\cdot)$ are univariate non-central chi-squared probability function and density function respectively.[10] $F(\cdot)$ has accurate approximations and can be obtained from IMSL. Therefore the bivariate probability can be computed with a one-dimensional numerical integration which is fast.

[EXAMPLE 2]
Using the parameter values as before, we compute the discount bond option prices for both Vasicek and CIR models. The option which is written on a 5-year bond expires in 6 months and the strike is 0.6 for $1 face value. The option prices are 0.1088 for the Vasicek model and 0.1085 for the CIR model. The non-central chi-squared probabilities are obtained from CSNDF in IMSL of FORTRAN.

Time-dependent parameter models will have closed form solutions for their option prices only if the underlying interest rate follows a normal process. If the underlying interest rate follows a non-normal process like the square root process, there will have no easy solution to the option. Because the bond pricing equation is an exponential form, under the normality assumption of the underlying interest rate, the future bond price should be log-normally distributed. Therefore, the classical Black–Scholes/Merton methodology can be used to find the option price. It is interesting to note that the bond itself needs not have a closed form equation. Recall that the bond

[10] Details can be found in Chen and Scott (1992).

solution is a general exponential function of the interest rate:

$$P(t,T) = A(t,T)e^{-rB(t,T)}.$$

It is not always the case that $A(\cdot,\cdot)$ and $B(\cdot,\cdot)$ have closed expressions. Since A and B contain no variable, it will not change the distribution of the bond price, P. As a result the bond price is always normally distributed. Its volatility is of course always:

$$\text{var}[\ln P(T,s)] = B(T,s)^2 \text{ var}[r(s)].$$

Note that although there should be a functional form for the volatility, as having been argued, this functional form needs not be known if one adopts the implied volatility from the previously traded option. The solution of the option will take exactly the same form of the Black–Scholes model except that the volatility of the underlying bond is not known.

Options on discount bonds are not common. What are common are futures options and coupon bond options. Treasure bond futures options and Eurodollar futures options are two very liquid exchange traded option contracts. Coupon bond options, on the other hand, are normally embedded in other contracts. The refinance option in mortgages is an important coupon bond option. The call provision in corporate and Treasury bonds is another example.

There is no simple way to value coupon bond option. However, under the single-factor framework, Jamshidian (1989) shows that a coupon bond option can be priced as if a portfolio of discount bond options. It should be noted that this decomposition works only under the single-factor framework. It will break down if the model contains more than one factor.

The option on a coupon bond has the payoff at maturity as follows:

$$O_c(T_c, T_c, \underline{s}) = \max\{Q(T_c, \underline{s}) - K, 0\}$$

where $\underline{s} = \{T_1, T_2, \cdots, T_n\}$ is a vector of coupon arrival times and the coupon bond price can be written as the weighted sum of a series of pure discount bonds:

$$Q(T_c, \underline{s}) = \sum_{i=1}^{n} c_i P(T_c, T_i)$$

where all c's are equal to the coupon amount except for the last c which is equal to the coupon plus the face value. We first write the strike price K in the same form as the bond as:

$$K = \sum_{i=1}^{n} c_i K_i$$

Substituting these two expressions back to the payoff function, we get:

$$\max\left\{\sum_{i=1}^{n} c_i\left[P(T_c, T_i) - K_i\right], 0\right\}.$$

Write out explicitly the solution of the bond price and duplicate the same form for each strike to get:

$$\sum_{i=1}^{n} c_i\left[A(T_c, T_i)e^{-r(T_c)B(T_c, T_i)} - A(T_c, T_i)e^{-r^* B(T_c, T_i)}\right].$$

It is important to note that once the form of each strike is made the same as the bond, whether or not an individual discount bond is greater than its corresponding strike depends on if the interest rate at time T_c is smaller than the "strike rate" r^*. It is then quite easy to see that if one particular bond, say $P(T_c, T_3)$, is greater than its strike K_3, then it indicates that $r(T_c)$ must be smaller than r^*, given that exponential functions are all monotonic. But if $r(T_c)$ is smaller than r^*, then it indicates that all discount bonds must be greater than their corresponding strikes, again for the monotonicity reason. This says that if one bond is greater (or less) than its strike, all bonds will be greater (or less) than their strikes. This immediate means that selective exercise has no value and the max function can be eliminated. Then the option can be treated as a weighted sum of a series of discount bond options:

$$\begin{aligned} O_c(t, T_c, \underline{s}) &= \hat{E}_t\left[\exp\left(-\int_t^{T_c} r(u)du\right)\max\left\{\sum_{i=1}^{n} c_i P(T_c, T_i) - K_i, 0\right\}\right] \\ &= \sum_{i=1}^{n} \hat{E}_t\left[\exp\left(-\int_t^{T_c} r(u)du\right)\max\left\{c_i P(T_c, T_i) - K_i, 0\right\}\right] \\ &= \sum_{i=1}^{n} c_i O_c(t, T_c, T_i) \end{aligned}$$

The trick for this approach to work is to find all the K's, or r^*. To find r^*, we need to perform a search for r so that it generates the coupon bond price that is equal to K. Then this r^* is plugged into each K equation to find all the K's. Finally, we use all the strikes to identify all the discount bond options and the weighted sum of them gives the coupon bond option value.

It is clear that when we extend the term structure to multiple factors, the trick will not work. It is impossible to find any single value for r under which both the bond and the strike can be expressed. We obtain a linear function of the state variables, instead.

3.3 BOND FUTURES OPTION

Bond futures option differs from the straight bond option in that futures contracts have no value. It is impossible for the holder of a futures option contract to use the strike price to exchange for a worthless futures contract. As a consequence, the futures option contract will have to be cash settled. The holder gets the difference between the futures price at the option expiry and the strike price, or 0, whichever is larger. At the valuation standpoint, the solution is different due to the different boundary condition for futures options. The call option will have the terminal payoff function $\max\{\Phi(T_c, T_f, T) - K, 0\}$. Therefore, the solution is:

$$
\begin{aligned}
O_{vas}(t) &= \hat{E}\left[\exp\left(-\int_t^{T_c} r(s)ds\right)\max\{\Phi(T_c, T_f, T) - K, 0\}\right] \\
&= \hat{E}\left[\exp\left(-\int_t^{T_c} r(s)ds\right)\Phi(T_c; T_f, T)\right]\bar{E}\left[1_{\{\Phi>K\}}\right] - \hat{E}\left[\exp\left(-\int_t^{T_c} r(s)ds\right)\right]K\bar{E}\left[1_{\{\Phi>K\}}\right] \\
&= H(t)\,N(d_h) - P(t, T_c)\,K\,N(d_h - \sigma_h)
\end{aligned}
$$

where

$$
\begin{aligned}
d_h &= \frac{1}{\sigma_h}\left(\ln\frac{H(t)}{P(t, T_c)\,K} + \frac{\sigma_h^2}{2}\right) \\
\sigma_h^2 &= \mathrm{var}[\ln H(T_c)] = X(T_c, T_f, T)^2\,\frac{\sigma^2\left(1 - e^{-2\alpha(T_c - t)}\right)}{2\alpha}.
\end{aligned}
$$

Again, this solution is the same for both one-factor and two-factor term structure models. It is very interesting to see that in the futures option pricing equation, the underlying asset is not a futures price, but becomes a hypothetical asset, H. This H asset pays the futures price at the expiration of the option but it is not equal to the futures price today. This is because futures contracts have no value. When an option is written on the futures contract, it is actually written on an asset that pays off the futures price. And this asset will earn riskless rate at the risk-neutral world. In other words,

$$
\begin{aligned}
H(T_c) &= \Phi(T_c; T_f, T) \\
H(t) &= \hat{E}\left[\exp\left(-\int_t^{T_c} r(s)ds\right)\Phi(T_c; T_f, T)\right]
\end{aligned}
$$

Note that H is not a traded asset, therefore, this formula needs parameters. Compared to the Black model (1976) for futures options, the function H is an expected discounted futures price. In the Black model, since the interest rates are assumed constant, discounting is trivial in contrast to the H function here where stochastic interest rates are considered. With the change of measure, the function H can be decomposed into:

$$\hat{E}_t\left[\exp\left(-\int_t^{T_c} r(s)ds\right)\Phi(T_c, T_f, T)\right] = P(t, T_c)\bar{E}_t\left[\Phi(T_c, T_f, T)\right].$$

As will be shown in the next chapter, the second term is a forward price of a contract where a futures contract is delivered at maturity, T_C. Note that H is preference-free:

$$H(t)_{vas} = P((t, T_c)\ \Phi(t, T_f, T)\ e^{-Z}$$

where

$$Z = \frac{\sigma^2}{2} F(t, T_c)^2 X(T_c).$$

The function $X(\cdot)$ is defined the Vasicek's futures formula. It is clear that the preference parameter drops out of the equation. They are embedded explicitly in the first term, the bond price, and implicitly the second term, the futures price, which contains two bond prices.

The futures option formula under the CIR framework for a one factor model can be derived similarly; only that the H function and the two probability functions need to be different:

$$\begin{aligned} H(t)_{cir} &= P(t, T_c)\bar{E}_t\left[\Phi(T_c, T_f, T)\right] \\ &= P(t, T_c)\bar{E}_t\left[C(T_c, T_f, T)e^{-r(T_c)D(T_c, T_f, T)}\right] \\ &= P(t, T_c)\bar{E}_t\left[C(T_c, T_f, T)e^{-x(T_c)D(T_c, T_f, T)/[2(\phi+\psi)]}\right] \\ &= P(t, T_c)C(T_c, T_f, T)\exp\left(\frac{-D(T_c, T_f, T)\Lambda^* /[2(\phi+\psi)]}{1+D(T_c, T_f, T)/(\phi+\psi)}\right)\left[1 + D(T_c, T_f, T) / (\phi + \psi)\right]^{-2\kappa\theta/v^2}. \end{aligned}$$

In this derivation, we use the forward adjusted process on the square root interest rate. The forward adjusted process for the square root model is when $x = 2(\phi+\psi)r$ where is a non-central chi-squared variate. The chi-squared probability functions are more complicated. We should just present the final results. These results were simultaneously developed by Jamshidian (1987) and Feldman (1993):

$$O_{cir}(t) = H_{cir}(t)\chi^2\left(2r^*(\phi + \psi + D(T_c, T_f, T)), \frac{4\kappa\theta}{\sigma^2}, \frac{2\phi^2 e^{\gamma(T_c - t)}}{\phi + \psi + D(T_c, T_f, T)}\right)$$
$$-P(t, T_c)K\chi^2\left(2r^*(\phi + \psi), \frac{4\kappa\theta}{\sigma^2}, \frac{2\phi^2 e^{\gamma(T_c - t)}}{\phi + \psi}\right).$$

Note that this formula is very similar to the bond option formula of CIR. The only difference is the asset price is now H. The two probabilities change only slightly. This is because the futures pricing function is a substitute for the bond pricing formula. We have observed this result in the Vasicek case in which the volatility is based upon the asset price of H.

Again, the futures option is solved by Chen and Scott (1992). The result is:

$$O(t) = H_1 H_2 \chi^2\left(\{y_1, y_2\}^*, \frac{4\kappa_1\theta_1}{\upsilon_1^2}, \frac{4\kappa_2\theta_2}{\upsilon_2^2}, \Lambda_1^*, \Lambda_2^*\right) - P_1 P_2 K \chi^2\left(\{y_1, y_2\}^\circ, \frac{4\kappa_1\theta_1}{\upsilon_1^2}, \frac{4\kappa_2\theta_2}{\upsilon_2^2}, \Lambda_1^\circ, \Lambda_2^\circ\right)$$

where definitions of the arguments can be found in the bond option formula except that B is replaced by D.

3.4 BOND FORWARD OPTION

Options on forward are similar to those on futures. The solution can be derived by substituting the forward price for the futures price. For a normal process like Vasicek, we obtain:

$$O_\Psi(t) = L(t)\, N(d_l) - P(t, T_c)\, K\, N(d_l - \sigma_l)$$

where

$$d_l = \frac{1}{\sigma_l}\left(\ln \frac{L(t)}{P(t, T_c)\, K} + \frac{\sigma_l^2}{2}\right)$$
$$\sigma_l^2 = \mathrm{var}[\ln L(T_c)] = \sigma_h^2.$$

The L function is also a preference-free function:

$$L(t) = \hat{E}\left[\exp\left(-\int_t^{T_c} r(s)ds\right)\Psi(T_c, T_f, T)\right]$$
$$= \hat{E}\left[\exp\left(-\int_t^{T_c} r(s)ds\right)\right]\overline{E}[\Psi(T_c, T_f, T)].$$

It is interesting to note that the forward-adjusted expectation of the forward price is not equal to the forward price. The bracketed term resembles the bond option formula with the option maturing at T_f instead of T_C. This bracketed term is similar to Turnbull and Milne (1991, Theorem 2). Details can be found in Chen (1995).

3.5 CONCLUSION

Interest rate claims can be viewed as claims on a non-traded asset, the instantaneous rate, or a traded asset, the pure discount bond. Therefore, interest rate claims can be preference-free or non-preference-free, depending upon which asset is chosen as the underlying state variable. In this chapter, it is shown that since bond prices are themselves contingent claims, the preference-free format of other claims (especially options) may or may not exist. If the bond price follows a known distribution, then the claim prices based upon bond prices are possible. Since bonds are traded assets, the preference-free format can be obtained. On the other hand, if the bond price distribution is not known (such as CIR), solutions can only be expressed only in terms of the instantaneous rate. The preference-free form will not be reachable.

The Vasicek two-factor model provides an interesting insight. If the bond price is taken from the market place as given, then it makes no difference whether the bond is priced by the one-factor model, the two-factor model, or the time-dependent model. The key ingredient lies in the bond volatility. In other words, pricing an option accurately depends upon whether the bond volatility can be correctly specified. Dybvig (1989) has provided excellent discussions on the theoretical development of this issue. The two-factor model certainly gives the volatility a much broader base.

Finally, it can be implied easily in the paper that one can do an n-factor decomposition of the instantaneous rate and the option formulas will again remain the same. The underlying assets (bond, forward, and futures) will only have to be moderately modified.[11]

APPENDIX

A. Method of Separation of Variables

Solving any partial differential equation requires guessing a form of the solution. If a specific form can solve the partial differential equation, then this must be *the* solution

[11] Sharp (1987, Chapter 5) shows n-factor bond and bond option models

according to the existence and uniqueness theorem in differential equations.

A solution is guessed to take the following form:

$$P(t,T) = e^{-rF(t,T)-G(t,T)}.$$

Then, all the partial derivatives will therefore become:

$$\frac{\partial P}{\partial r} = P \cdot (-F)$$
$$\frac{\partial^2 P}{\partial r^2} = P \cdot F^2$$
$$\frac{\partial P}{\partial t} = P \cdot \left(-r\frac{\partial F}{\partial t} - -\frac{\partial G}{\partial t}\right)$$

Plugging these partials back into the differential equation, one can obtain three separate ordinary differential equations to solve:

$$\alpha F - \frac{dF}{dt} - 1 = 0$$
$$\frac{dG}{dt} + (\alpha\mu - q\sigma)F - \frac{\sigma^2}{2}F^2 = 0$$

All of them are the first order ordinary differential equation in t which are easy to solve. Solving the first two ordinary differential equations for F and using them to solve for G in the third ordinary differential equation, one will obtain a homogeneous solution. The original boundary condition, $P(T,T\} = 1$, can be translated into conditions for Fs and G as follows

$$F(T) = G(T) = 0.$$

This will help to obtain the particular solution.

B. Change of Probability Measure

This appendix presents the second changes of measure, i.e., $\hat{E}_t\left[e^{-\int r}P\right] = \hat{E}_t\left[e^{-\int r}\right]\overline{E}_t[P]$ and $\hat{E}_t\left[e^{-\int r}1_{\{P>K\}}\right] = \hat{E}_t\left[e^{-\int r}\right]\overline{E}_t\left[1_{\{P>K\}}\right]$. For the change of measure of a two-factor model, see Chen (1995). By Ito's rule:

$$\begin{aligned}
0 &= \ln P(T_c, T_c) \\
&= \ln P(t, T_c) + \int_t^{T_c} \tfrac{1}{P}\left(P_u du + P_r dr + \tfrac{1}{2} P_{rr}(dr)^2\right) \\
&= \ln P(t, T_c) + \int_t^{T_c} \tfrac{1}{P}\left(P_u + \alpha(\mu - r - \tfrac{\sigma q}{\alpha})P_r + \tfrac{\sigma^2}{2} P_{rr}\right)du + \int_t^{T_c} \tfrac{1}{P}\sigma P_r d\hat{W}(u) - \int_t^{T_c} \tfrac{1}{2}\left(\sigma \tfrac{P_r}{P}\right)^2 du \\
&= \ln P(t, T_c) + \int_t^{T_c} r(u)du + \int_t^{T_c} \tfrac{1}{P}\sigma P_r d\hat{W}(u) - \int_t^{T_c} \tfrac{1}{2}\left(\sigma \tfrac{P_r}{P}\right)^2 du
\end{aligned}$$

Re-arranging terms, we get:

$$-\ln P(t, T_c) - \int_t^{T_c} r(u)du = \int_t^{T_c} \sigma \tfrac{P_r}{P} d\hat{W}(u) - \int_t^{T_c} \tfrac{1}{2}\left(\sigma \tfrac{P_r}{P}\right)^2 du$$

$$\frac{\exp\left(-\int_t^{T_c} r(u)du\right)}{P(t, T_c)} = \exp\left(\int_t^{T_c} \sigma \tfrac{P_r}{P} d\hat{W}(u) - \int_t^{T_c} \tfrac{1}{2}\left(\sigma \tfrac{P_r}{P}\right)^2 du\right)$$

This proves the second change of measure that:

$$\begin{aligned}
\eta_2 &= \frac{e^{-\int_t^{T_c} r(u)du}}{P(t, T_c)} \\
&= \exp\left[\int_t^{T_c} \sigma \frac{P_r}{P} d\hat{W}(u) - \int_t^{T_c} \tfrac{1}{2}\left(\sigma \frac{P_r}{P}\right)^2\right] \\
&= \frac{d\overline{\wp}}{d\hat{\wp}}
\end{aligned}$$

The first change of probability measure goes through the same process:

$$\ln P(T_c, T) = \ln P(t, T_c) + \int_t^{T_c} r(u)\, du + \int_t^{T_c} \sigma \tfrac{P_r}{P} d\hat{W}(u) - \int_t^{T_c} \tfrac{1}{2}\left(\sigma \tfrac{P_r}{P}\right)^2 du.$$

Therefore,

$$\frac{d\tilde{\wp}}{d\hat{\wp}} = \eta_1$$

C. Using the Black–Scholes Result

Since the change of measure in the Black–Scholes formula can apply to all log-norm[al] processes, it can be applied in our options. Let any state variable $S(t)$ follow a risk-neutralize[d] log-normal process that satisfies the stochastic differential equation:

$$dS = rS\,dt + \sigma S\,d\hat{W}$$

where r is the risk-free rate and s is the volatility parameter. Then the call option formula can be written as the following conditional expectation:

$$\begin{aligned}
C &= e^{-r(T-t)}\hat{E}[\max\{S_T - K, 0\}] \\
&= e^{-r(T-t)}\hat{E}[(S_T - K)\,1_{\{S>K\}}] \\
&= e^{-r(T-t)}\hat{E}[S_T\,1_{\{S>K\}}] - e^{-r(T-t)}K\,\hat{E}[1_{\{S>K\}}] \\
&= e^{-r(T-t)}\hat{E}[S_T]\,\hat{E}\left[\frac{S_T}{\hat{E}[S_T]}\,1_{\{S>K\}}\right] - e^{-r(T-t)}K\,N(d_2) \\
&= S_t\,\hat{E}[\eta\,1_{\{S>K\}}] - e^{-r(T-t)}K\,N(d_2) \\
&= S_t\,\overline{E}[1_{\{S>K\}}] - e^{-r(T-t)}K\,N(d_2) \\
&= S_t\,N(d_1) - e^{-r(T-t)}K\,N(d_2)
\end{aligned}$$

where $N(d_2)$ is a normal probability defined upon the process $dS = rS\,dt + \sigma S\,d\hat{W}$ and therefore $d_2 = [\ln(S/K) + (r - \sigma^2/2)(T-t)]/\sigma\sqrt{T-t}$. Now, write:

$$\begin{aligned}
&\ln S_T = \ln S_t + \int_t^T \left(r - \tfrac{\sigma^2}{2}\right)du + \int_t^T \sigma\,d\hat{W} \\
&\ln S_T - \ln S_t - \int_t^T r\,du = \int_t^T \sigma\,d\hat{W} - \int_t^T \tfrac{\sigma^2}{2}\,du \\
&\frac{S_T}{S_t e^{r(T-t)}} = \frac{S_T}{\hat{E}[S_T]} = \eta
\end{aligned}$$

This means $N(d_1)$ is defined upon $dS = (r+\sigma)S\,dt + \sigma S\,d\overline{W}$ and therefore verifies that $d_1 = d_2 + \sigma\sqrt{T-t}$. This gives equation (12) an alternative derivation:

$$\begin{aligned}
&\hat{E}\left[\exp\left(-\int_t^{T_c} r(s)ds\right) P(T_c,T)\,1_{\{P>K\}}\right] \\
&= \hat{E}\left[\exp\left(-\int_t^{T_c} r(s)ds\right)\right]\tilde{E}\left[P(T_c,T)\,1_{\{P>K\}}\right] \\
&= P(t,T_c)\,\tilde{E}[P(T_c,T)]\,\tilde{E}\left[\frac{P(T_c,T)}{\tilde{E}[P(T_c,T)]}\,1_{\{P>K\}}\right] \\
&= P(t,T_c)\cdot\frac{P(t,T)}{P(t,T_c)}\cdot\tilde{E}\left[\frac{P(T_c,T)}{\tilde{E}[P(T_c,T)]}\,1_{\{P>K\}}\right] \\
&= P(t,T)\,\tilde{E}\left[\frac{P(T_c,T)}{\tilde{E}[P(T_c,T)]}\,1_{\{P>K\}}\right] \\
&= P(t,T)\,\tilde{E}\left[\eta\,1_{\{P>K\}}\right]
\end{aligned}$$

Note that P follows a log-normal process. Hence, the previous result can be applied directly and will result in $P(t,T)N(d_1)$ and $V_p = \mathrm{var}(\ln P(T_c,T)$. This approach is convenient because any variable that follows a log-normal process will have this result. As a consequence, both futures option and forward option can be derived using the above procedure because both H and L follow log-normal processes.

CHAPTER 4
COMMON INTEREST RATE CONTRACTS

There are several well known and high volume interest rate contracts. Exchange traded Eurodollar futures and futures options, Treasury bond futures and futures options, short term interest rate options, as well as over the counter swaps, caps and floors are all important interest rate contracts. MBSes recently have also gained tremendous attention. Their pricing therefore poses an important question. The basic pricing of these contracts relies on the materials in the previous chapter while various considerations need to be given to each contract's uniqueness.

4.1 EURODOLLAR FUTURES, OPTIONS, AND SHORT-TERM INTEREST RATE OPTIONS

In this section, for simplicity, we use only a one-factor Vasicek model to demonstrate the pricing issues of each contract. Other models can be derived in a similar manner.

4.1.1 Eurodollar Futures

Eurodollar futures are futures contracts settled on LIBOR (London InterBank Offer Rate). A three month LIBOR is an money market rate that can be translated into a discount bond:

$$L = 4\left(\frac{1}{P(t, t+.25)} - 1\right).$$

A Eurodollar futures is a futures contract that at the expiration of the futures contract (T_f), a Eurodollar time deposit of $1 million with a three-month maturity will be used as the futures payoff. There are some confusion to be clarified about Eurodollar futures. First, the quote seen in the newspaper is not a quote of a discount bond price but 1–LIBOR. For example, a quote of 92 means the futures price will be settled at a LIBOR rate of 1–0.92=8%. Then this 8% is treated as a discount yield to define a settlement price of the futures, which is 1–8%/4=0.98. For $1 million face value, the payoff is therefore $980,000. Secondly, Eurodollar futures are cash settled. Since LIBOR is a money market rate and not a discount yield, the settlement price is not a discount bond price. As a result, the futures contract is a contract on LIBOR. Using the pricing formula in the previous chapter, we can write the current Eurodollar futures price as:

$$\Phi(t) = \hat{E}_t[1 - L / 4].$$

Converting LIBOR to a discount bond price as above, we have:

$$\Phi(t) = \hat{E}_t\left[2 - \frac{1}{P(T_f, T_f + 0.25)}\right].$$

If we take the Vasicek formula, then the solution is easy:

$$\begin{aligned}\Phi(t) &= 2 - \hat{E}_t\left[e^{+r(T_f)F(T_f,T)+G(T_f,T)}\right] \\ &= 2 - e^{G(T_f,T)+F(T_f,T)\hat{E}_t[r(T_f)]+F(T_f,T)^2 V_t[r(T_f)]/2}\end{aligned}$$

and the mean and variance of the rate are known and given in Chapter 2. If we take the CIR formula, then:

$$\begin{aligned}\Phi(t) &= 2 - \hat{E}_t\left[A(T_f, T)e^{+r(T_f)B(T_f,T)}\right] \\ &= 2 - A(T_f, T)\frac{\exp\left(\frac{2cB(T_f,T)\lambda}{1+4cB(T_f,T)}\right)}{(1 - 4cB(T_f, T))^{4\kappa\theta/\upsilon^2}}\end{aligned}$$

Note that in pricing Eurodollar futures, we implicitly assume an instantaneous riskless rate underlying LIBOR so that we can use the discount bond futures pricing formula. This is of course inappropriate because LIBOR is an interbank rate that carries credit risks. LIBOR usually lies above the T Bill rate for this very reason. More careful modeling needs to be used. However, for simplicity, in practice people assume the LIBOR rate is a Vasicek solution so that some parameters of the interest rates can come in and explain the futures price. In other words, often people set:

$$\begin{aligned}\Phi(t) &= 1 - \hat{E}_t[L] / 4 \\ &= 1 - e^{\hat{E}_t[\ln L]+\hat{V}_t[\ln L]/2} / 4\end{aligned}$$

for a three-month LIBOR.

4.1.2 Eurodollar Futures Options

If we take the last futures pricing formula for Eurodollar futures, then the option pricing formula is equally easy. Taking a risk-neutral expectation of the futures formula:

$$\begin{aligned} C(t) &= \hat{E}_t\left[\exp\left(\int_t^{T_c} r(u)du\right)\max\left\{1-\tfrac{1}{4}\Phi(t,T_f,T_f+.25)-K,0\right\}\right] \\ &= \tfrac{1}{4}\hat{E}_t\left[\exp\left(\int_t^{T_c} r(u)du\right)\max\left\{K*-\Phi(t,T_f,T_f+.25),0\right\}\right] \end{aligned}$$

where $K^* = 4(1-K)$. It is clear that Eurodollar futures call becomes a put option on the pure discount futures contract. The formula in Chapter 2 can be readily used. A more correct formula based upon the correct setting of the LIBOR definition can also be derived, but we leave that to the readers.

4.1.3 Short-Term Interest Rate Option

CBOT trades some light volume of short term interest rate options. These options are written discount rates, so they are really options on discount bonds. The bond option formula in Chapter 2 can be directly applied here. Note the quote of a discount rate is the following definition:

$$R = (1-P(t,T))/(T-t).$$

The option value is the risk-neutral expectation of the payoff and is:

$$\begin{aligned} C(t) &= \hat{E}_t\left[\exp\left(\int_t^{T_c} r(u)du\right)\max\{R-K,0\}\right] \\ &= \hat{E}_t\left[\exp\left(\int_t^{T_c} r(u)du\right)\max\left\{[1-P(T_c,T)]/(T-T_c)-K,0\right\}\right] \\ &= \hat{E}_t\left[\exp\left(\int_t^{T_c} r(u)du\right)\max\left\{K*-P(T_c,T),0\right\}\right]/(T-T_c) \end{aligned}$$

where $K^* = 1-K(T-T_c)$. This valuation equation is once again the direct application of the put option result of the previous chapter. Note that the option on a rate is turned into an option on a bond. However, we can use the same assumption as in the case of Eurodollar futures options where the underlying rate, R in this case, is assumed to be log-normally distributed. Then, the solution is precisely a Vasicek bond option formula.

4.2 TREASURY BOND FUTURES AND THE QUALITY OPTION

4.2.1 Straight Treasury Bond Futures

Straight T Bond futures are easy to handle. Recall that we have easy solutions for discount bond futures in both Vasicek and CIR versions. T Bonds are coupon bonds. By the arbitrage argument in Chapter 1, we know that we can use the $1 discount bonds at different maturities as discount factors to discount coupons. Therefore, we can write the coupon bond formula as:

$$Q(t, \underline{s}) = \Sigma_{j=1}^{n} c \cdot P(t, T_j) + 100 \cdot P(t, T_n)$$

where c is the per $100 coupon payment. We use $\underline{s}$ as the collection of all maturities from 1 to n.

The futures price is simple. Recall the risk-neutral expectation we have used all along. The futures price is the risk-neutral expectation of the future bond price. Since now the bond price is Q, we shall just take an expected value of Q in the future time:

$$\begin{aligned}
\Phi(t, T_f, \underline{s}) &= \hat{E}_t\left[Q(T_f, \underline{s})\right] \\
&= \hat{E}_t\left[\Sigma_{j=1}^{n} c \cdot P(T_f, T_j) + 100 \cdot P(T_f, T_n)\right] \\
&= \Sigma_{j=1}^{n} c \cdot \hat{E}_t[P(T_f, T_j)] + 100 \cdot \hat{E}_t[P(T_f, T_n)] \\
&= \Sigma_{j=1}^{n} c \cdot \Phi(t, T_f, T_j) + 100 \cdot \Phi(t, T_f, T_n)
\end{aligned}$$

We can see that the coupon bond futures price is a weighted sum of discount bond futures prices. Again, once the parameters of the model are known, this coupon bond futures price is easily calculated. T Bond futures can use precisely this formula. Unfortunately, T Bond futures have embedded options. The most important option is the quality option. There are timing options that are also important.

4.2.2 The Delivery Options

A T Bond futures contract requires the short position to physically deliver the underlying asset, any Treasury bond that has at least 15 years to maturity or first call, upon expiration. Because of the flexibility, the short can choose the cheapest bond to deliver. This is known as the quality option in T-Bond futures. Although the CBOT has designed conversion factors to compensate the long position of the futures contract by asking the short to increase the quantity of the underlying asset, if the short chooses a lesser quality asset (or vice versa), this pre-determined factor cannot completely eliminate the quality option. Historically, the futures price has still been below the price implied by the cost of carry model, reflecting the non-trivial quality option value.

Delivery of a T-Bond futures contract can occur at any time in the last month of trading. This is called the delivery month. There is a 3-day period for delivery as follows: (1) the position day when the short decides to deliver, (2) the notice day when a long position is assigned delivery and notified, and (3) the delivery day when the transaction is settled. The first delivery day is the first business day of the delivery month and the last delivery day is the last business day of the delivery month.

The timing option in T-Bond futures is the ability for the short to make delivery any time in the delivery month.[12] Due to different trading hours of the T-Bond market and the T-Bond futures market, the timing option can occur at three different times. A regular timing option is the ability to make the delivery any time when both markets open. The timing option also allows the short to still make delivery even when the futures market is closed. First, the end of month timing option occurs because the last trading day is 8 business days from the end of the delivery month. The last trading day is therefore 7 business days before the last delivery day. Allow for 2-day settlement, that is if a short position at the close of the last trading day buys a bond in the spot market and notifies the exchange of a delivery, the transaction is settled 2 business days later, which is 5 business days before the last delivery day, the end of month timing option covers 6 business days, or roughly 1 week.

The futures market at the Chicago Board of Trade (CBOT) opens at 8:20 a.m. Eastern time and closes at 3:00 p.m. Eastern time but the spot market opens till 9:00 p.m. Eastern time. Therefore, the short has 6 hours to decide if he wants to deliver at the settlement price. This is typically called the wild card timing option.

The quality option has been widely studied by researchers.[13] A simple arbitrage argument can show that the quality option can be viewed as an option on the minimum. Stulz (1982) and Johnson (1987) have analyzed such an option and provide closed form solutions in a form of multi-dimensional normal integrals if the underlying assets

[12] The timing option needs a numerical method which is not discussed in this chapter. See Boyle (1989) for an excellent discussion of the timing option.

[13] See Chance and Hemler (1993) for an excellent review.

follow a joint log normal process. Since the multi-dimensional integral cannot be obtained without numerical methods and numerical methods can directly apply to option calculations, multi-dimensional integrals do not provide any computational efficiency. As a result, most empirical studies assume two deliverable bonds so that the univariate Margrabe's exchange (1978) option formula can be used.[14]

Among all the delivery option studies, recent studies directly on the quality option of the T-Bond futures have been Kane and Marcus (1984), Benninga and Smirlock (1985), Hedge (1988), and Hemler (1990). Among them only Hemler uses a closed-form solution but his solution derives from a log-normal process for the bond price which lacks the support of the term structure theory. It is not until Carr (1988) that a model for the quality option consistent with the current observed term structure has been made available. Carr recognized that, ignoring the timing option, the futures contract with the quality option is the minimum of all the deliverable bonds. The solution, in order to be consistent with the term structure theory, is based upon a random process for the interest rate, not a log-normal process for the bond price. He claims that a Black–Scholes type of closed form solution can be identified.

4.2.3 Models for T Bond Futures

Given the existence of the quality option, at the maturity of the futures contract, the short can choose the lowest quality bond to deliver. The bond chosen is called the cheapest-to-deliver bond. The payoff the short position receives is the futures price times the conversion factor minus the cheapest bond price. In formula, this means the payoff is:

$$\max\left\{q_1 f - Q(T_f, \underline{T}_1), q_2 f - Q(T_f, \underline{T}_2), \cdots, q_n f - Q(T_f, \underline{T}_n)\right\} = 0.$$

To avoid arbitrage, the futures price must be set to 0 so that the short cannot gain any profit. To solve for this equation, we shall arrive at the following solution to the futures price at delivery:

$$\Phi_{TB}(T_f) = \min\left\{\frac{Q(T_f, \underline{T}_1)}{q_1}, \frac{Q(T_f, \underline{T}_2)}{q_2}, \cdots, \frac{Q(T_f, \underline{T}_n)}{q_n}\right\}$$

[14] Because in this case, Stulz's bivariate integral will reduce to a univariate integral. See Hemler (1990) and his citation of the previous work using the Margrabe formula.

where $\Phi_{TB}(\cdot)$ is the futures price at maturity with n eligible bonds to deliver, each of which is represented by $Q(\cdot)$ and q's are conversion factors.

For simplicity, let there be only two deliverable bonds, then the solution is quite simple and a Black–Scholes type of solution can be identified. At the expiration, the payoff of such futures contract simplifies to:

$$\Phi_{TB}(T_f) = \min\left\{\frac{Q(T_f, \underline{T}_1)}{q_1}, \frac{Q(T_f, \underline{T}_2)}{q_2}\right\}.$$

It is then necessary to determine regions where one bond is cheaper and regions where the other is cheaper. It will be shown later on that over a plausible range of interest rates, there exists only one crossover rate that separates a real line into two non-overlapping regions. In this case, the current futures price can be solved easily:

$$\begin{aligned}
\Phi_{TB}(t) &= \hat{E}_t\left[\min\left\{\frac{Q(T_f, \underline{T}_1)}{q_1}, \frac{Q(T_f, \underline{T}_2)}{q_2}\right\}\right] \\
&= \int_{-\infty}^{\infty} \min\left\{\frac{Q(T_f, \underline{T}_1)}{q_1}, \frac{Q(T_f, \underline{T}_2)}{q_2}\right\}\varphi(r)dr \\
&= \int_{-\infty}^{r^*} \frac{Q(T_f, \underline{T}_1)}{q_1}\varphi(r)dr + \int_{r^*}^{\infty} \frac{Q(T_f, \underline{T}_2)}{q_2}\varphi(r)dr.
\end{aligned}$$

An explicit solution can be arrived at if an explicit assumption of the interest rate is assumed, for example, to be Vasicek or CIR. If it is a one-factor Vasicek, then the solution is easy:

$$\begin{aligned}
\Phi_{TB}(t) &= \Sigma_{j=1}^{J_1} \frac{c_{1j}}{q_1}\Phi(t, T_f, T_{1,j})N\left(\frac{r^* - m - F(T_f, T_{1j})V}{\sqrt{V}}\right) \\
&+ \Sigma_{j=1}^{J_2} \frac{c_{2j}}{q_2}\Phi(t, T_f, T_{2j})N\left(-\frac{r^* - m - F(T_f, T_{2j})V}{\sqrt{V}}\right)
\end{aligned}$$

where

$$m = r(t)e^{-\alpha(T_f - t)} + \mu\left(1 - e^{-\alpha(T_f - t)}\right)$$

$$V = \frac{\sigma^2\left(1 - e^{-2\alpha(T_f - t)}\right)}{2\alpha}$$

If r is assumed to follow CIR, the solution is:

$$\Phi_{TB}(t) = \Sigma_{j=1}^{J_1} \frac{C_{1j}}{q_1}\Phi(t, T_f, T_j)\chi_j^2(r^*) + \Sigma_{j=1}^{J_1} \frac{C_{1j}}{q_1}\Phi(t, T_f, T_j)[1 - \chi_j^2(r^*)]$$

where

$$\chi_j^2(r^*) = n.c.\chi^2\left(2(\eta + B(T_j))r^*; \frac{4\kappa\theta}{v^2}, \frac{2\eta^2 e^{-(\kappa+\lambda)T_f} r}{\eta + B(T_j)}\right),$$

$$\Phi(t, T_f, T) = C(t, T_f, T)e^{-rD(t, T_f, T)},$$

$$C(t, T_f, T) = \left(\frac{\eta}{\eta + B(t,T)}\right)^{2\kappa\theta/v^2} A(T_f, T),$$

$$D(t, T_f, T) = \frac{\eta e^{-(\kappa+\lambda)(T_f - t)}}{\eta + B(t,T)} B(T_f, T),$$

and

$$\eta(t, T_f) = \frac{2(\kappa+\lambda)}{v^2(1 - e^{-(\kappa+\lambda)(T_f - t)})}.$$

Note that $\Phi(\cdot,\cdot,\cdot)$ is the discount bond futures price for the Vasicek or the CIR model. For more than two bonds, the solution can be identified similarly. Interested readers can refer to the original article by Carr (1988).

Timing options are much more complex to analyze and need numerical methods. There are three timing options: regular timing option, daily wild card option, and the end of month timing option. A detailed analysis of the timing options can be found in Carr, Chen, and Scott (1995).

4.2.4 Two-Factor Models

As pointed out earlier, the one-factor term structure model is not sufficient to generate a yield curve matching what has been observed. The one-factor model suffers from the fact that all bonds are perfectly correlated and therefore the flexibility of the yield curve is limited. This disadvantage is more of a problem in long term bonds than short-term bonds. Since all bonds that can qualify for delivery need to be 15 years or longer, the one-factor model is not appropriate. In fact, using the parameters in Chen and Scott (1993) for the one-factor model, we cannot find any crossover rate for all the delivery dates in the 87–91 period, suggesting that one-factor model severely misprices the long term bonds.

Using a two-factor model would cause a technical difficulty. There exists an area where the option is in the money, not a line. Therefore, in search for the crossover rate, one needs to specify a number of combinations of the two factors.

The two-factor model suggests the following solution for the futures price:

$$\begin{aligned}\Phi(T_f) &= \int_0^\infty \int_0^\infty \min\left\{\frac{Q(T_{b1})}{q_1}, \frac{Q(T_{b2})}{q_2}\right\}\varphi(y_1, y_2)dy_1dy_2 \\ &= \iint\limits_{y_1+y_2\in A} \frac{Q(T_{b1})}{q_1}\varphi(y_1, y_2)dy_1dy_2 + \iint\limits_{y_1+y_2\in A^c} \frac{Q(T_{b2})}{q_2}\varphi(y_1, y_2)dy_1dy_2 \\ &= \iint\limits_{y_1+y_2\in A} \frac{Q(T_{b1})}{q_1}\varphi(y_1, y_2)dy_1dy_2 + \iint\limits_{y_1+y_2\in A^c} \frac{Q(T_{b2})}{q_2}\varphi(y_1, y_2)dy_1dy_2.\end{aligned}$$

This solution is obviously not in closed form and therefore not easy to evaluate. However, with simple probability functions, the solution can be approximated with good accuracy.

4.3 SWAPS

In general, an interest rate swap (with no embedded options) is the exchange of two series of coupon cash flows, which can be either floating-floating or fixed-floating. Most interest rate swap contracts including currency swaps can be characterized as one of the two kinds. The fixed-floating swap is generally a par swap which involves no cash changing hands at inception. The pricing of such a swap is to determine the fixed rate. In the floating-floating swap one counterparty has to pay the other a premium. The pricing of such a swap requires the calculation of the premium.

The pricing of swaps has long been widely studied among academics and practitioners. Because plain vanilla swaps can be priced and hedged easily,[15] most swap pricing studies focus on embedded options in swaps. Most common embedded options in swaps are the default option and the cancellation option. Default option normally is viewed as the credit risk of the counterparty to the swap. Cancellation option is viewed as the interest rate risk of the swap. Early swap pricing literature focuses mostly on the default option. Whittaker (1987) treats the termination of an interest rate swap (by the fixed rate payer) as a joint decision to exercise the call option to buy back the fixed rate debt and the put option to sell back to the floating rate payer the floating rate debt. It is shown in this paper that such a treatment of the default option is not appropriate because the swap cannot be simply splitted into a call and a put options. Sundaresan (1991) derives an interest rate swap model under the square root term structure. To reach an easy closed form solution for the default option, he borrows a framework of Merton (1976) and models the default risk premium directly

[15] Bicksler and Chen (1986) argue that a vanilla interest rate swap can be viewed as fixed rate debt exchange for floating rate debt.

instead of a state variable underlying the default events. This formulation of the default option cannot determine the level of the premium. The default option model by Cooper and Mello (1991) characterizes default as company insolvency and therefore is priced by the standard Black–Scholes methodology. Although their model allows stochastic interest rates, it cannot be used to price interest rate swaps with both default and cancellation options.

4.3.1 The Framework

Historically, fixed-floating swaps occupy the largest portion of the swap market and can be viewed as fixed-rate debts swapping for floating-rate debts. However, due to credit and interest rate risks caused by defaults and cancellations, this view cannot hold as a valid pricing concept. One has to pursue more sophisticated machinery in order to price swaps with these risks.

In general, the floating rate cannot be the instantaneous short rate. The common example of the floating rate is LIBOR which is a money market rate. A vanilla swap is a swap contract where no default or cancellation risk is considered. Both counterparties have to honor the contract and finish their cash flows. The pricing of such type swaps is very straightforward. Note that for every exchange of cash flows, the fixed leg (fixed rate payer) pays a fixed coupon for a random amount (floating coupon). This is like a long forward contract to the fixed leg of the swap where a fixed forward price is paid to acquire the underlying spot which is worth a random amount. Since this is true for every exchange of cash flows, a swap is simply a collection of forward contracts. Therefore, pricing such a swap is to set the fixed coupon rate (forward price) so that no money changed hands at inception. Since a vanilla swap contains a series of such forward contracts, we have to set one forward price (i.e., fixed coupon rate) to all the forward contracts maturing at various times over the life of the swap. It is then necessary to define a forward rate as follows:

$$f(t,s) = -\frac{d\ln P(t,T)}{dT}$$

where $\Psi(\cdot)$ is the forward price maturing at time s of a pure discount bond maturing at time T. This forward rate is the same as the forward rate in Section 1.3 only that this forward rate is in continuous time. If the bond process is represented as follows:

$$\frac{dP}{P} = \mu_p(P,t)dt + \sigma_p(P,t)dW.$$

then the forward rate can be computed by the following expectation of the future spot rate:

$$f(t,s) = \overline{E}_t[r(s)].$$

The proof of the result is given in an appendix. This provides another interpretation of the linkage between the forward rate process used by Heath, Jarrow, and Morton (1992) and the spot rate process (forward-adjusted).

As mentioned earlier, a plain vanilla swap is nothing but a series of forward contracts. As a result, the pricing of a swap contract is just the recovery of the forward price. Since each forward contract can be decomposed into a long call and a short put, the vanilla swap can be thought of as a collection of calls and puts. To solve this forward pricing problem in our framework, we set the risk-neutral expectation to 0 and solve for the fixed coupon rate:

$$0 = \hat{E}_t\left[\sum_{i=1}^{n} e^{-\int_t^{t_i} r(u)du}(r(t_i) - c)\right]$$

$$0 = \sum_{i=1}^{n} \hat{E}_t\left[e^{-\int_t^{t_i} r(u)du}(r(t_i) - c)\right]$$

$$0 = \sum_{i=1}^{n} P(t,t_i)\overline{E}_t[r(t_i) - c].$$

The last line of the above equation is a result of change of probability measure. The derivation of such change of measure can be found in an appendix. This change of measure will give a probability space that defines the forward-adjusted process. The original derivation of the forward-adjusted process can be found in Jamshidian (1987, 1989). Simplifying the solution, we can arrive at the solution to the fixed coupon rate:

$$c = \sum_{i=1}^{n} w_i\overline{E}_t[r(t_i)]$$

where $w_i = \frac{P_i}{\sum_i P_i}$.

The fixed coupon now is the weighted average of forward rates. In a single period swap (n=1), it is intuitive that the fixed coupon rate should equal the forward rate, as indicated in the equation. Given the current term structure of discount bonds, if forward rates are observable, then there is no need for further calculations. If forward rates are not observable, then we need to compute each forward rate as a forward-adjusted expectation of the instantaneous short rate so that the fixed coupon rate in the vanilla swap can be priced accordingly.

4.3.2 Option to Cancel

Cancellation defined in this section is introduced only through interest rate fluctuations. Contract termination due to defaults are treated separately in which we assume defaults occur due to non-interest-rate events such as technical insolvancy. The treatment of the cancellations (or interest rate risk) requires the use of the compound option methodology. Smith, Smithson, and Wakeman (1987) first proposed the use the compound option model of Geske (1977) to price swaps under credit risk. However, they do not provide any form of the pricing model to be used in conjunction with the term structure theory. To see in our framework that a swap with the option to cancel is a compound put option, we start with the payoff at the end of maturity. If the floating leg (floating rate payer) of a fixed-floating swap is allowed to cancel the contract, then the payoff for the fixed leg (fixed rate payer) is:

$$\min\{r(t_n)-c,\ 0\} \text{ or } -\max\{c-r(t_n),\ 0\}$$

which is a short put option. Obviously, the floating leg will cancel the contract if the floating rate is higher than the fixed rate. Similarly, if the fixed leg owns a similar option to cancel, then the cash flow received by the floating leg is:

$$\min\{c-r(t_n),\ 0\} \text{ or } -\max\{r(t_n)-c,\ 0\}$$

which is a short call option for the floating leg (a long call to the fixed leg.) However, it is not correct to state conclude that the fixed leg of the swap owns a call and shorts a put. The valuation of the cancellation cannot be applied to both sides simultaneously. This is because if both counterparties are allowed to hold the option to cancel, then there would be no settlement at expiration, since one of the counterparties will surely cancel the contract.

Assume that the floating leg is the only counterparty that has the option, then at the period prior to maturity (continue to assume the floating leg to have the option),

$$\min\left\{r(t_{n-1}) - c + \hat{E}_{t_{n-1}}\left[e^{-\int_{t_{n-1}}^{t_n} r(u)du} \min\{r(t_n) - c,\ 0\}\right],\ 0\right\}.$$

Use $\zeta(t_{n-1})$ to represent the put option:

$$\hat{E}_{t_{n-1}}\left[e^{-\int_{t_{n-1}}^{t_n} r(u)du} \max\{c - r(t_n),\ 0\}\right]$$

and rewrite the equation as follows:

$$\min\{r(t_{n-1}) - c - \zeta(t_{n-1}),\ 0\}.$$

Clearly, this short put option will worsen (more negative) the payoff at t_{n-1} for the fixed leg. The floating leg will not cancel the contract if the interest rate level is only higher than c, unlike the last period. Because the floating leg owns a put option, he will not cancel the contract unless the interest rate is higher than $c + \zeta(t_{n-1})$. This slightly more reluctance to cancel results from the reason that the option in the next period may become in the money. This is clearly a case of compound option discussed by Geske (1977). As this compound put option accumulates in value, the floating leg is more and more unlikely to cancel the contract. On the other hand, the fixed leg would like to set the fixed rate, c, as low as possible so that the put option that it sells to the floating leg has a minimal value (lower strikes produce lower put premiums). In terms of pricing where the fixed leg needs to choose a 'fair' fixed rate so that the swap is transacted at par, any non-zero put premium is a cost to the fixed leg and therefore the fixed leg should ask for a lower fixed rate to compensate for the put it sells to the floating leg. As a result, par swaps are not possible under this compound option case. The fixed leg would need to charge a premium for this put option. Working backwards, we can arrive at the current swap value as:

$$\zeta(t) = P(t, t_n) \sum_{i=1}^{n} \int_0^{c+\zeta(t_1)} \cdots \int_0^{c+\zeta(t_i)} (r(t_i) - c)\overline{\varphi}_i \prod_{j=1}^{i} dr_j$$

where $\overline{\varphi}_i = \overline{\varphi}(r(t_1), \cdots, r(t_i) | r(t))$ is an i-dimensional density function of the interest rates.

This solution differs from the standard compound put option formula in that in each period, there is an interest rate variable in the payoff function. Obviously, the compound option solution is not in closed-form except for two-period, normally or log-normally distributed underlying state variable. With $n>2$ periods, there exists no fast approximation formulas that can calculate the option value.[16] In an appendix, we show how to derive this equation.

In reality, cancellations rarely exist because in general swap contracts will include a penalty if the out-of-money leg decides to unwind the position. For example, a 5-year swap having a swap rate of 6% is agreed by two counterparties. Three years later, the 2-year swap rate rises to 8%. The fixed leg that pays 6% of the 5-year contract can engage in a short 2-year swap and receive the difference of 2%. For the fixed leg of the 5-year swap, the contract is in the money. The out-of-money floating leg can now unwind the position but it needs to compensate the fixed leg the present value of 2% for the remaining 2 years. This payment works as a penalty against cancellation. As a result, there is no incentive for the floating leg to cancel the contract.

4.3.3 Option to Default

In this section, we confine defaults as purely non-interest-rate events and therefore can be diversified away. If the counterparty has a strong financial record, it is regarded as low credit risk because it is unlikely to bankrupt which would lead to the default of the swap contract. Early work of the credit risk can be traced to Merton's (1976) jump processes. If the counterparty, due to its financial difficulties, has to discontinue the swap, the swap will be forced to be cancelled. We model default events with a Poisson jump process and not treat default as an insolvency result. As a result, default in our model will not have an option-like payoff. This is because in this paper we view a swap as a financial security; not the total debt of a corporation. Therefore, we price swaps as we price other traded financial securities.

Let the default process be modeled by a Poisson process with parameter λ. Then, by Merton (1976), we have an expected cash flow for a given future time:

$$e^{-\lambda(t_i - t)}(r(t_i) - c).$$

To have a par swap, we simply set the coupon rate so that the initial value of the swap is

---

[16] Even in the simplest normality case, the multi-variate normal probabilities require simulations which are no faster than Monte Carlo simulations or the lattice method to calculate option values directly.

0:

$$0=\sum_{i=1}^{n} e^{-\lambda(t_i-t)}\hat{E}_t\left[e^{-\int_t^{t_i} r(u)du}\left(r(t_i)-c\right)\right]$$
$$=\sum_{i=1}^{n} e^{-\lambda(t_i-t)}P(t,t_i)\left(\overline{E}_t[r(t_i)]-c\right).$$

As a consequence, the fixed coupon rate can be set as:

$$c=\frac{\sum_{i=1}^{n} e^{-\lambda(t_i-t)}P(t,t_i)\overline{E}_t[r(t_i)]}{\sum_{i=1}^{n} e^{-\lambda(t_i-t)}P(t,t_i)}=\sum_{i=1}^{n} w_i\overline{E}_t[r(t_i)]$$

where $w_i=\dfrac{e^{-\lambda(t_i-t)}P_i}{\sum_i e^{-\lambda(t_i-t)}P_i}$.

This is another weighted average result. The only difference between this result and that of the vanilla swap is that the numerator of each weight is lessened by a probability. It can be seen that under this default risk, early weights are higher than those in the vanilla case. As the default likelihood becomes higher (λ increases), early forward rates are weighted more and future forward rates are less. In the limiting case where $\lambda=\infty$, all weights are concentrated on the first forward rate. The intuition is that when the default is likely to happen, the fixed leg must pay less attention to distant coupons and therefore will weight near forward rates more heavily than distant ones. When $\lambda = 0$, i.e., no default, all weights are identical to those in the vanilla case.

4.3.4 Cancellation and Default

It has been seen that the option to cancel is a compound option but a reverse swap or a penalty payment should negate the option value while a default option is not an option because defaults occur involuntarily and they can happen to both counterparties. What has been found to be an interesting question is that when both issues are considered. When a default occurs, the defaulting counterparty will default on the swap if it is out of the money but will not default if it is in the money. Therefore, for the surviving counterparty, the payoff of a swap is no longer vanilla but will look like the following:

$$\min\{0,-Q(\tau,\underline{s})\}=-\max\{0,Q(\tau,\underline{s})\}$$

where t is a random default time of the swap. Since this option involves random stopping time, the solution is not easily available. One needs to use numerical methods.

4.3.5 Swaptions

Swaptions are exchange-traded interest rate options. Holders of swaptions have rights to purchase (become the fixed leg) a swap at a fixed time with a fixed swap rate. If the underlying swap is a vanilla type, then the valuation of such options is straightforward. Note that the payoff of a swaption is:

$$\max\{0, Q(T_f, \underline{s}) - K)$$

where

$$Q(T_f, \underline{s}) = \Sigma c_j(T_f)P(T_f, T_j)$$
$$K = \Sigma \bar{c}_j P(T_f, T_j)$$

Note that $c(T_f)$ is the market swap rate at the expiration which is greater than $\bar{c}$ which is strike swap rate. Therefore, the value of a swaption is merely an option on coupon bonds. The reason is that the swaption allows its holder to use a lower swap rate for the remaining life of the swap. This is like giving the holder a coupon bond of with the coupon equal to $c(T_f) - \bar{c}$.

As discussed in the previous chapter, the coupon bond option value can be easily found under the single-factor model. For multi-factor models, calculations are not easy because there is no single value of the interest rate that can make selective exercise worthless. To see that, we write a two-factor bond pricing model:

$$Q(t, \underline{s}) = \Sigma c_j e^{-y_1(t)F_1(t,T_j) - y_2(t)F_2(t,T_j) - G(t,T_j)}$$
$$K = \Sigma c_j e^{-\bar{y}_1(t)F_1(t,T_j) - \bar{y}_2(t)F_2(t,T_j) - G(t,T_j)}$$

There is no guarantee that if one bond is less than its strike, all other bonds will be less than their strikes. $P(t, T_j) > K_j$ implies:

$$[y_1(t) - \bar{y}_1(t)]F_1(t, T_j) + [y_2(t) - \bar{y}_2(t)]F_2(t, T_j) < 0$$

and there is no guarantee that for another bond, the value:

$$[y_1(t)-\bar{y}_1(t)]F_1(t,T_k)+[y_2(t)-\bar{y}_2(t)]F_2(t,T_k)$$

can still be negative. The valuation has been sketched by Chen and Scott (1992). Interested readers can refer to the original article for details.

4.3.6 Differential Swaps

Differential swaps represent a series of similar swap contracts in which both legs of the swap commit to floating interest rates. There are many names for many types of differential swaps such as rate differential swaps, LIBOR differential swaps, index swaps, cross rate swaps, and currency protected swaps. Although there are so many different types of differential swaps, they all share one common characteristic, which is both parties of the swap are committed to floating interest rates. In this section, we employ our methodology to exemplify one differential swap and other swaps can be priced with slight changes of the derivation used in this section.

Most differential swaps involve initial exchange of cash flows since there exists no fixed coupon rate. In this section, we shall define a differential swap in which one counterparty uses the domestic LIBOR to exchange for the foreign LIBOR. The settlement is done in domestic currency. Assuming no default, we can write the payoff of each period as a forward-adjusted expected value as desired above:

$$\bar{E}_t[L^*-L]=\bar{E}_t\left[\frac{1}{P^*(T_f,T_f+\Delta t)}-\frac{1}{P(T_f,T_f+\Delta t)}\right]\Big/\Delta t$$

where L^* and L are two LIBOR rates (foreign and domestic respectively) which lead to two discount bond prices $P^*(\cdot)$ and $P(\cdot)$ respectively in foreign and domestic countries. This calculation is sufficiently more sophisticated because now we need to characterize the interest rate behavior in two countries and the correlation between them. To demonstrate, we use a joint OU process for the two instantaneous rates underlying LIBORs. Let:

$$\begin{bmatrix} dr^* \\ dr \end{bmatrix}=\begin{bmatrix} \alpha_1(\mu_1-r^*) \\ \alpha_2(\mu_2-r) \end{bmatrix}dt+\begin{bmatrix} \sigma_1 & 0 \\ 0 & \sigma_2 \end{bmatrix}\begin{bmatrix} dW_1 \\ dW_2 \end{bmatrix}$$

where $E[dW_r dW_{r^*}]=\rho dt$. Through orthogonization, we can rewrite the joint process in a form that obtaining a forward adjusted process can be made:

$$\begin{bmatrix} dr^* \\ dr \end{bmatrix} = \begin{bmatrix} \alpha_1(\mu_1 - r^*) \\ \alpha_2(\mu_2 - r)dt \end{bmatrix} dt + \begin{bmatrix} \sigma_1\sqrt{1-\rho^2} & \rho\sigma_1 \\ 0 & \sigma_2 \end{bmatrix} \begin{bmatrix} dZ_1 \\ dZ_2 \end{bmatrix}.$$

The forward adjusted process can therefore be written as:

$$\begin{bmatrix} dr^* \\ dr \end{bmatrix} = \begin{bmatrix} \alpha_1(\mu_1 - r^*) - \sigma_1 q_1 - \rho\sigma_1\sigma_2 P_r / P \\ \alpha_2(\mu_2 - r)dt - \sigma_2 q_2 - \sigma_2 P_r / P \end{bmatrix} dt + \begin{bmatrix} \sigma_1\sqrt{1-\rho^2} & \rho\sigma_1 \\ 0 & \sigma_2 \end{bmatrix} \begin{bmatrix} d\bar{Z}_1 \\ d\bar{Z}_2 \end{bmatrix}.$$

Note that P^* which is the domestic price of a foreign bond needs be calculated first and then its forward price. As one can see these calculations involve both processes. Once the exchange rate is another variable, the calculations will become even more complex.

Currency swaps are in general special forms of differential swaps. Thus, no matter how complicated a currency swap, by changing differential swaps slightly, one can easily price currency swaps. In this section, for only illustration purposes, we price only the simplest swap which involves only one state variable.

Assume a German company wants to invest in the US and therefore it needs US dollars to begin its investments. Instead of borrowing from a US bank, it can swap with a US company with a similar interest in Germany. A simplest swap can be arranged as follows. If the current exchange rate is X and both parties decide to swap $\$A$ million dollars worth of currencies. Then the German company will pay the US company B $(=A/X)$ million marks for US$\$A$ million. Further assume that the exchanges of interests are fixed for fixed, say US$\$c_1 A$ million (German company pays US company) for $c_2 B$ million marks (US company pays German company) where c_1 and c_2 are two fixed coupon rates. At expiration, the German company will transform its investment US$ A million back into marks and pay the US company B million marks. This is the simplest form of currency swap, called fixed–fixed vanilla currency swap. For the German company, its net cash flow at each period is its inflow minus its outflow (in millions):

$$c_2 BX - c_1 A$$

If only the exchange rate is random, this is a simple fixed–floating swap. If both exchange rate and coupon rates are floating, then this is a differential swap. Certainly, the valuation of such swaps is not simple. However, with numerical methods, all types of swaps are solvable.

4.3.7 A Simple Practice — Credit Exposure

We have learned that once defaults are introduced into the swap, there exists no simple solutions, even to the simplest fixed–floating swap; not to mention complex variations. In order to have a simple measure of the option value. Wall street firms have developed a concept of "credit exposure". Credit exposure tells that when two firms carry different credit/default risks, say AAA versus BB, then one can quantify this credit difference (often called credit spread) into basis points. The good-rating company will worry its counterparty might default on its swap and therefore be exposed under the credit risk. Hull (1993) has demonstrated that such a credit exposure is a call option. To see that, if the swap is out of money for the good-rating firm, then the company is subject to no risk. If the swap is out of money for the bad-rating firm, then the company is subject to some credit risk because once the bad-rating company defaults, the company collects no money from the swap. As a consequence, the credit exposure for the company is:

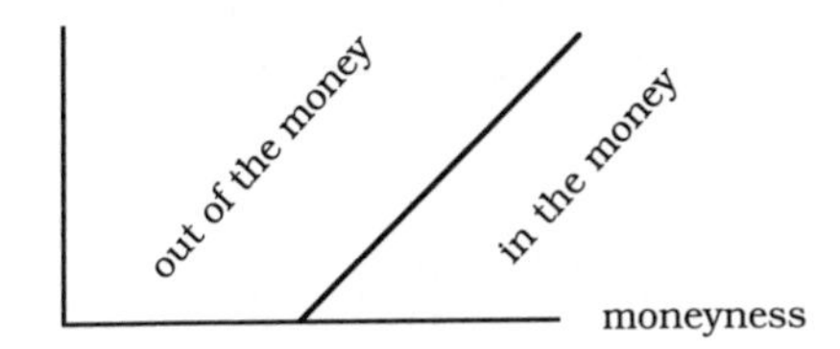

As an approximation, this exposure is given to each coupon swap date. If we further assume that the floating rate is log-normally distributed, then the value of the credit exposure can be calculated by the Black–Scholes formula. This assumption also allows us to derive alternative formulas. To conserve space, we simply list three interesting results under log normal, normal, and square root assumptions.

OU PROCESS

$$\overline{E}_t[\max\{r(t_i)-c,\ 0\}]$$
$$=\overline{E}_t[r(t_i)]\Phi\left(\frac{\overline{E}_t[r(t_i)]-c}{\overline{\upsilon}_r}\right)+\overline{\sigma}_r\phi\left(\frac{\overline{E}_t[r(t_i)]-c}{\overline{\upsilon}_r}\right)-c\cdot\Phi\left(\frac{\overline{E}_t[r(t_i)]-c}{\overline{\upsilon}_r}\right)$$

where $\Phi(\cdot)$ is a standard normal probability function, $\phi(\cdot)$ is a standard normal density function and $\overline{\upsilon}_r=\sqrt{\overline{V}_t[r(s)]}=\sqrt{\frac{\delta^2\left(1-e^{-2\alpha(s-t)}\right)}{2\alpha}}$.

SR PROCESS

$$
\begin{aligned}
&\overline{E}_t[\max\{r(t_i)-c,\ 0\}] \\
&= \frac{\Lambda}{2(\phi+\psi)}\chi^2(2(\phi+\psi)c;\Delta+4,\Lambda)+\frac{\Delta}{2(\phi+\psi)}\chi^2(2(\phi+\psi)c;\Delta+2,\Lambda)-c\cdot\chi^2(2(\phi+\psi)c;\Delta,\Lambda) \\
&= \overline{E}_t[r(t_i)]\chi^2(2(\phi+\psi)c;\Delta+4,\Lambda)+\frac{\Delta}{2(\phi+\psi)}\left[\frac{[(\phi+\psi)c]^{(\Delta/2+1)}e^{-(\phi+\psi)c}}{\Gamma(\Delta/2+2)}\right]-c\cdot\chi^2(2(\phi+\psi)c;\Delta,\Lambda)
\end{aligned}
$$

where $\Lambda = r(t)\frac{2\phi^2 e^{\gamma(s-t)}}{\phi+\psi}$ represents the magnitude of non-centrality and $\Delta = \frac{4\kappa\theta}{\upsilon^2}$ is the degrees of freedom. The proof of this result which uses a recursive formula of the Bessel function can be obtained from the author on request.

LN PROCESS

$$
\begin{aligned}
&\overline{E}_t[\max\{r(t_i)-c,\ 0\}] \\
&= e^{\overline{\mu}_r+\frac{\overline{\sigma}_r^2}{2}}\Phi\left(\frac{c-\overline{\mu}_r}{\overline{\sigma}_r}+\overline{\sigma}_r\right)-c\cdot\Phi\left(\frac{c-\overline{\mu}_r}{\overline{\sigma}_r}\right) \\
&= \overline{E}_t[r(t_i)]\Phi(d+\overline{\sigma}_r)-c\cdot\Phi(d)
\end{aligned}
$$

where

$$
d = \frac{\ln(\overline{E}_t[r(t_i)]/c)-\eta^2(t_i-t)/2}{\eta\sqrt{t_i-t}},
$$

$$
\overline{\mu}_r = \overline{E}_t[\ln r(t_i)] = \ln \overline{E}_t[r(t_i)] - \frac{\overline{V}[\ln r(t_i)]}{2}, \text{ and}
$$

$$
\overline{\sigma}_r^2 = \overline{V}_t[\ln r(t_i)] = \eta^2(t_i-t).
$$

4.4 CAPS AND FLOORS

Caps and floors are common interest rate options. Generally, caps and floors are embedded in floating rate instruments to protect investors (floors) or issuers (caps). For a floating rate note, a cap is a ceiling interest rate beyond which the company pays only the cap rate. A floor in a floating rate note gives the lower bound of the interest rate payment so that investors are protected from low interest rates. Caps and floors can also be bought separately to accompany straight floating rate instruments. In this case, a premium would be paid. If caps and floors are embedded in the floating rate note, then instead of charging a premium, the issuer will commonly convert the premium into a mark-up (caps) or mark-down (floors) to adjust the index interest rate.

4.4.1 Black–Scholes Valuation

Caps and floors are options. Take the cap for example, it is a call option on the rate or put option on a bond. The Black–Scholes formulas for calls and puts are generally used as an approximation. With interest rate models, more sophisticated pricing formulas can be derived. To see that the cap is an option, look at the coupon payoff of a floating rate note:

$$\begin{aligned}\tau L \min\{R, \overline{R}\} &= \tau L(R + \min\{0, \overline{R} - R\}) \\ &= \tau L(R - \max\{0, R - \overline{R}\})\end{aligned}$$

where τ is the time till coupon arrival and L represents the principal. It is seen from the above formulation that a cap in a floating rate bond is a straight floating rate bond with a call option written on it. Therefore the cap value is the value of the call option. If R is log normally distributed, then the famous Black–Scholes formula can be used to value such a cap. In fact, practitioners do use the Black–Scholes formula for cap valuation. Let us look at the following example.

[EXAMPLE 1]
Suppose a one year floating rate with a face value of $1M pays semi annual coupons with the index rate being 6 month LIBOR. Today's LIBOR is 6% and the cap rate is set to be 7%. If the volatility (σ, standard deviation of log of LIBOR) is 10%, then what is the cap value? The answer is $335. The details of this solution is given as follows.

The coupons to be received are one certain amount in 6 months and one uncertain amount in one year. The one in 6 months needs no option valuation. It is the one in one year that needs option valuation because we do not know if the LIBOR rate will or will not exceed the cap rate. The payoffs are drawn in the following time line:

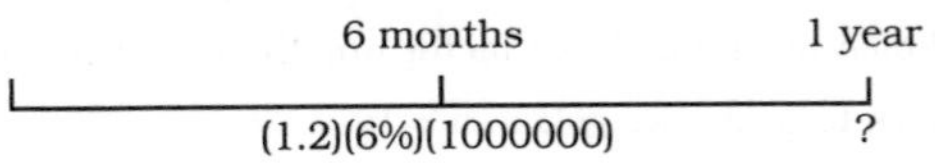

	6 months	1 year
	(1.2)(6%)(1000000)	?

The present value of the cap in one year is the Black–Scholes result:

$$C = R_0 N(d_1) - \tfrac{1}{1+y} \overline{R} N(d_2)$$

where y is the one year riskless yield and $N(\cdot)$ is a standard normal probability. In this case, we assume y is 7%. The two arguments in the normal probability functions are defined similar to those in the standard Black–Scholes formula. With numbers plugged in, we can find the value of the cap:

$$\begin{cases} d_1 = \left[\ln \frac{R_0}{\bar{R}/(1+y)} + \frac{\sigma^2 T}{2}\right] \Big/ \sigma\sqrt{T} = \left[\ln \frac{6\%}{7\%/(1+7\%)} + \frac{1\%}{2}\right] \Big/ \sigma\sqrt{T} = -0.81492 \\ d_2 = d_1 - \sigma\sqrt{T} = -0.81492 - 0.1 = -0.91492 \end{cases}.$$

The two normal probabilities are therefore equal to 0.21 and 0.18. Substituting these two numbers into the call formula to get:

$$\begin{aligned} C &= (6\%)(0.21) - \tfrac{1}{1.07}(7\%)(0.18) \\ &= 6.7 \text{ bps} \end{aligned}$$

which, in terms of the dollar value, is 0.5×1000000×0.00067=$335. In other words, if this cap is sold separately from the floating rate bond, the value is $335. The issuer needs to pay the investor $335 for acquiring the cap. In general, the cap is embedded in the floating rate bond and, for the convenience of trading, issuer will not pay the investor the cap value. Instead, the issuer will compensate the investor by increasing the coupon. The value of the incremental coupon should be equal to the cap value. To convert the cap value into a coupon increment, we set the new coupon rate as LIBOR plus x. The incremental coupons are therefore (0.5)(x)(1000000) in 6 months and in one year. Discounting these coupons (note that the 6 month yield is R_0 or 6% and the one year yield is assumed to be 7%), we get the present value of:

$$\begin{aligned} &\frac{1000000 \times x}{2}\left(\frac{1}{1+6\%/2} + \frac{1}{1+7\%}\right) = 335 = \frac{1000000 \times 6.7}{2} \\ &x = 3.52 \text{ bps.} \end{aligned}$$

Some texts use the Black formula for futures options to price caps. This is the same as using the Black–Scholes formula. If we factor our the one year discount factor, $\frac{1}{1+y}$, from the Black–Scholes formula, we get:

$$C = \tfrac{1}{1+y}\left[(1+y)R_0 N(d_1) - \bar{R}N(d_2)\right]$$

where conceptually, $(1+y)R_0$ if treating R_0 an asset, is a forward rate $f(0,1,1.5)$. Therefore, the Black formula instead of the Black–Scholes formula can be used. In our opinion, which formula to use depends on which rate is easier to acquire in the marketplace.

The floor can be obtained in a similar manner. Notice that the payoff of a floored floating rate bond is:

$$\tau L \max\{R, \underline{R}\} = \tau L\left(R + \max\{0, \underline{R} - R\}\right).$$

Therefore, the floor is a put option on LIBOR. Using the same information in the example and the Black–Scholes put formula, we can calculate the floor value as follows (for the floor rate being 5%):

$$\begin{aligned} \tfrac{1}{1+y}\underline{R}N(-d_2) - R_0 N(-d_1) &= \tfrac{1}{1.07}(5\%)(0.007147) - (6\%)(0.005389) \\ &= 0.11 \text{ bps.} \end{aligned}$$

Also similar to the cap, the floor value can be translated into a markdown for the issuer since it is an option held by the investor.

If the issuer is issuing both cap and floor for the same floating rate bond, in general, the issuer will set the cap and the floor rates so that the two option values cancel each other. In this case, the two values cancel if the floor rate is set at 5.91%.

If a floating rate bond has more than one year till maturity, then the cap or floor will have more than one option. The total value of a cap or a floor is therefore the sum of all option values. Although more options are involved, the calculation of the markup and markdown remains the same.

4.4.2 Valuation Under an Interest Rate Model

Caps and floors are interest rate options. Interest rate options differ from equity options in that inputs of the option valuation formula need to be consistent with term structure observed from the market place. In particular, the volatility structure needs to be consistent with the market. It is well known that in the Black–Scholes formula, the volatility of the asset is linear in time, i.e., the volatility (variance) is proportion to the time to maturity. However, for bonds, volatility will first increase with time because uncertainty is greater distant bonds. As bonds get close to maturity and the face value, the volatility will decline. Therefore, the volatility structure of a bond is a humped curve. If this volatility structure is not considered in pricing caps and floors, we believe that the options will be significantly mispriced.

To price caps and floors under a term structure theory, we first need to convert the call option of a cap into a put option of a bond. To see that,

$$
\begin{aligned}
\max\{R-\overline{R},0\} &= \max\left\{2\left(\frac{1}{P(0,0.5)}-1\right)-\overline{R},0\right\} \\
&= \max\left\{\frac{2}{P(0,0.5)}-(2+\overline{R}),0\right\} \\
&= \frac{2+\overline{R}}{P(0,0.5)}\max\{K-P(0,0.5),0\}
\end{aligned}
$$

where $K = 2/(2+\overline{R})$. It is interesting to note that initially a cap is a call option on LIBOR but now it becomes a put option on a pure discount bond. If we further assume that the LIBOR is driven by a single-state variable, the instantaneous short rate and it follows the Vasicek model, then the Jamshidian's put option formula can be used. Similarly, the put option of a floor will become a call on a discount bond.

Other than replacing the Black–Scholes formula with the Jamshidian formula, all analyses remain the same. We should note that the single factor Vasicek model cannot explain the observed volatility structure in the market. As a result, more complex models such as multi factor or time dependent models should be used.

[EXAMPLE 2]

Using parameters values for the Vasicek model given in Chapter 2, we calculate the cap value of the above example using the interest rate option formula. The cap rate is set to be 8.5%. The results are 10.43 bps for the one year cap.

4.5 MORTGAGE-BACKED SECURITIES

Mortgage backed securities are like interest rate securities because mortgages are just like bonds; without the face value. Therefore, mortgage backed securities can be treated as interest rate contracts and the analyses of the interest rate contracts can be readily applied to mortgage-backed securities. However, there is one catch. Mortgages can have defaults and prepayments!

Mortgages defaults are just like bond defaults and the analyses of corporate bonds can be applied. Prepayments, on the other hand, can be the big problem. In general, prepayments are categorized into three areas. They are prepayments caused by refinance, by non-economical reasons such as divorcing, moving, and others, and by irrational behaviors. The refinance prepayments are interest rate options. The homeowner has the right to use the new loan to pay back the old loan with a lower mortgage rate. The non-economical prepayments cannot be modeled by any interest rate models and are usually treated with a pure white noise. The irrational prepayments are very difficult to handle. Not only do they present no pattern but also they tend to contaminate the other two prepayments. Irrational prepayments are people who should not refinance, refinance; or those who should refinance, do not. In general, they are caused by the ignorance of the market or lack of economic sense. A lot of people tend to prepay when they have idle cash because it is thought to be traditionally bad to owe money. Because mortgages are so difficult to handle, interest rate models alone cannot meet the needs. So far, there have been no reasonable models for mortgage backed securities. Investment banks still rely heavily on the simple net present value method to value most of their mortgage-backed securities.

In this section, we should focus the interest risk that MBSes have. In other words, we shall pretend that all prepayments except for the refinance do not exist. We shall talk about the basic mortgage first and then several MBS contracts, such as Mortgage bonds, CMOs, IOs, and POs.

4.5.1 A Mortgage

The refinance option in a mortgage goes as follows. The homeowner has the right to prepay the old loan with the new loan at a lower mortgage rate. We know that when a mortgage is taken, the payment schedule is fixed through the end of the mortgage. This schedule is normally called the "amortization schedule".

[EXAMPLE 2]
The amortization schedule for a 15 year, $200,000 mortgage at 8% can be computed as follows. First, we shall determine its monthly payment. This can be solved by the annuity method. The monthly payment equals:

$$P = \frac{200,000}{PVIFA_{\overline{8\%/12|}180}} = 1911.30$$

In the first month, the interest is 200,000×8%/12=1333.33. Therefore, the principal paid is 1911.30–1333.33=577.97. For the next month, the new principal is 200000–577.97=19422.03. This new principal is used to compute the interest for the second month. Since the payment is fixed at 1911.30, the interest proportion is getting less and less and the principal paid is getting more and more. At the last month the principal is reduced to exactly 1911.30 and is paid off completely. Then the loan is terminated.

If the homeowner wants to prepay, he pays the running balance in the schedule, because that is the amount he still owes the mortgage company, usually a bank. Suppose the 15-year mortgage rate drops, say to 7% after 5 years. The running balance showing in the schedule should be 157,532.52 which is also the present value of 120 payments of 1911.30 discounted at the 8%/12 monthly rate. If the homeowner takes a new loan of 10 years (120 payments) at 7%, then he has to pay only 1829.09, which is 82.22 dollars less than before. So, he should definitely take a new loan and prepay the old loan. The value of his savings is 82.22×PVIFA(7%/12,120)=7081.14.

Another way to look at this problem is that when the mortgage rate drops to 7%, we can discount our 120 remaining payments of 1911.30 per month at the new rate. That gives 164,613.66, a higher value than the balance in the book which is 157,532.52. The difference is exactly 7081.14, the value of savings. Usually, we refer this to the value of the refinance option.

4.5.2 The Refinance Option Formula

To formulate the refinance option formally, we write at any time the homeowner's choice between the mortgage value, M, and the running balance in the book, B as:

$$\min\{B, M\}$$

where B is the book value and M is the new (market) mortgage value after the mortgage rate changes. When the mortgage rate drops, the homeowner should refinance and take a new loan. The new loan amount is B. So B should replace M as the market value. If the mortgage rate increases, then the mortgage value, M, drops. The homeowner should keep the mortgage because it is now valued at only M (while he should really owe $B > M$). In other words, the present value of all his future payments is only M.

Therefore, at any given time, we can write this payoff as:

$$\begin{aligned} &M + \min\{B - M, 0\} \\ &= M - \max\{M - B, 0\} \end{aligned}.$$

From the bank's point of view, this is like giving a mortgage without the ability to refinance and a call option to the borrower. This call option will reduce the value of the loan. In other words, if a homeowner wants to loan $200,000 and take the option to refinance in the future, then the bank will not be willing to loan him 200,000. It will have to be less than 200,000. The difference should reflect the value of the option. If the homeowner insists a loan of 200,000, then the company will charge a higher mortgage rate.

What should be the "right" rate? The right rate can be solved through an iterative process. Suppose the mortgage rate without the refinance option is 7%. Then, we can compute the call option value. This value should be used to reduce the loan amount that the company is willing to loan. To get exactly the same loan, we need to repeatedly substitute a new rate to the valuation equation until M is equal to the previous loan amount. Then this new rate is a mortgage rate with the refinance option.

The complexity comes in because the refinance option can be taken any time throughout the life of the mortgage. Therefore, the refinance option is an American option, not an European one. Since American options have no closed-form solutions, the refinance option needs to be solved numerically, say by the binomial method. And setting a mortgage rate for the loan with the refinance becomes a complex job.

4.5.3 Mortgage Bonds

Mortgages are pooled and sold in the secondary markets. When they are pooled, different maturities as well as different mortgage rates are pooled together. As a result, mortgage pools are not homogeneous. This non-homogeneity creates a lot of problems in pricing. In order to let investors understand what they invest in, pooling agencies normally restructure mortgage pools in into forms that are similar to existing instruments. For example, mortgage bonds are like bonds. Since mortgages receive fixed payments, it is quite natural to design a mortgage bond that gives its investors coupons.

Mortgage bonds are like bonds but subject to default risks. Therefore they are in general earning higher returns than T Bonds. Some mortgage bonds can carry a little prepayment risk. That is such bonds do not have fixed maturity dates.

4.5.4 Collateral Mortgage Obligations

One common restructure of mortgage pools is a horizontal cut. Each horizontal cut is called a tranche. The first tranche is entitled to the first few years of coupons. Once the first tranche retires, the second tranche takes over. The number of tranches can be as few as 3 or as many as 30. And each tranche's duration can also vary. Since there is prepayment risk in the pool, there is no guarantee that the later tranches can receive their coupons. So, later tranches suffer more prepayment risk than earlier tranches. In return, later tranches enjoy higher return. In some CMOs later tranches can earn as high as 300% rate of return.

CMO designs can vary, since they are all customer-taylored. There can be some risk reduction designs in some CMO tranches. A recent invention of PACs is such an example. A PAC, or Planned Amortization Class, guarantees its investors fixed coupons, making it very low prepayment risk. Interested readers can find descriptions of CMO's in professional books.

4.5.5 Interest Only and Principal Only

Some cuts make a pool pay its investors only interests or only principal, called IOs or POs. It is clear from the names that an IO receives only interests and a PO receives only the principal. It is also clear that when prepayment is high, IO investors will suffer because they will get less return. Since high prepayments occur when interest rates are low (high refinance activities), IO is considered to have "negative duration", i.e., lower interest rates, lower IO value. This is contrast to a regular bond that has positive duration, i.e., lower interest rates, higher bond price.

POs work like bonds and have positive duration. But POs have negative convexity because significant interest drops will not increase PO values more than slight interest rate drops.

4.5.6 Other Mortgage-Backed Contracts

Other mortgage-backed contracts are numerous. Direct mortgage passthroughs are the simplest MBSes. They are tiny slices of mortgage pools. Each slice is like a regular mortgage, only in small size. Servicing rights are fees charged by pooling agencies.

They are usually a fixed percentage of the pool value. If a pool is prepaid, then there is of course no charge. Therefore, servicing rights are like IOs that are functions of the prepayment speed.

APPENDIX

A. Derivation of the Forward Rate with Forward Measure i.e., $\overline{E}_t[r(s)] = f(t,s)$

The definition of the forward rate is:

$$f(t,s) = -\frac{d \ln P(t,s)}{ds}.$$

Substitute the risk-neutral valuation for the bond price and switch the differentiation and integration to arrive at the final expression as:

$$\begin{aligned} f(t,s) &= \frac{1}{P(t,s)} \frac{d}{ds} \hat{E}_t\left[e^{-\int_t^s r(u)du}\right] \\ &= \frac{1}{P(t,s)} \hat{E}_t\left[\frac{d}{ds} e^{-\int_t^s r(u)du}\right] \\ &= \frac{1}{P(t,s)} \hat{E}_t\left[e^{-\int_t^s r(u)du} r(s)\right]. \end{aligned}$$

From the appendix provided next, the risk-neutral expectation on the last line can be separated into two expectations, one under the risk-neutral process, the other under the forward-adjusted process Since the first expectation gives the bond price, the forward rate can be shown to be:

$$\begin{aligned} f(t,s) &= \frac{1}{P(t,s)} P(t,s)\overline{E}_t[r(s)] \\ &= \overline{E}_t[r(s)]. \end{aligned}$$

This completes the proof.

B. Change of Measure

Assume any claim function $G(r(T))$ of the interest rate. We need to show that:

$$\hat{E}_t\left[e^{-\int_t^T r(u)du} G(r(T))\right] = \hat{E}_t\left[e^{-\int_t^T r(u)du}\right]\overline{E}[G(r(T))] = P(t,T)\overline{E}[G(r(T))]$$

where $\overline{E}[\cdot]$ is taken under the forward-adjusted process. To prove this result, we first recognize that the forward rate expectation can be expressed as:

$$\overline{E}_t[G(r(T))] = \hat{E}_t\left[\frac{e^{-\int r}}{\hat{E}[e^{-\int r}]}G(r(T))\right]$$

Define $\xi = \frac{e^{-\int r}}{\hat{E}[e^{-\int r}]}$ as the Radon–Nikodym derivative. It is clear that $\hat{E}[\xi] = 1$. If we can find θ so that:

$$\xi = \exp\left(\int_t^T - \theta(u)dW(u) - \frac{1}{2}\int_0^T \theta^2(u)du\right).$$

Then we can identify a Girsanov transformation,

$$d\overline{W} = d\hat{W} + \theta(t)dt,$$

that turns the risk-neutral process into an equivalent process. We shall prove that this equivalent process is precisely the forward-adjusted process. Apply Ito's lemma to the log bond price:

$$\begin{aligned}
0 &= \ln P(T,T) \\
&= \ln P(t,T) + \int_t^T \frac{1}{P}\left[P_u + (\alpha - \lambda)P_r + \frac{\delta^2}{2}P_{rr}\right](r,u,T)du - \int_t^T \frac{1}{2}\left[\frac{\delta P_r}{P}\right]^2 (r,u,T)du \\
&\quad + \int_t^T \left[\frac{\delta P_r}{P}\right](r,u,T)d\hat{W}(u) \\
&= \ln P(t,T) + \int_t^T r(u)du - \int_t^T \frac{1}{2}\left[\frac{\delta P_r}{P}\right]^2 (r,u,T)du + \int_t^T \left[\frac{\delta P_r}{P}\right](r,u,T)d\hat{W}(u)
\end{aligned}$$

Rearranging terms and recognizing that the bond price is the risk-neutral expectation of the discount factor, we get:

$$-\ln P(t,T) - \int_t^T r(u)du = -\int_t^T \frac{1}{2}\left[\frac{\sigma P_r}{P}\right]^2 (r,u,T)du + \int_t^T \left[\frac{\sigma P_r}{P}\right](r,u,T)d\hat{W}(u)$$

$$\frac{e^{-\int_t^T r(u)du}}{\hat{E}\left[e^{-\int_t^T r(u)du}\right]} = -\int_t^T \frac{1}{2}\left[\frac{\sigma P_r}{P}\right]^2 (r,u,T)du + \int_t^T \left[\frac{\sigma P_r}{P}\right](r,u,T)d\hat{W}(u)$$

Given that $\xi = \frac{e^{-\int r}}{\hat{E}[e^{-\int r}]}$, we have identified θ to be $\sigma(r,t)P_r(t,T)/P(t,T)$. Therefore, we know the Girsanov transformation transforms the risk-neutralized process into a forward-adjusted process by changing the drift by $\sigma(r,t)\frac{\sigma(r,t)P_r(t,T)}{P(t,T)}$ amount.

C. The Compound Option Formula for Swaps

To make the notation simple, we denote the interest rate at time t_i as r_i instead of $r(t_i)$ used in the text. The interest rate at the current time is r_t. To derive the compound option formula for the swap with interest rate risk, we jump to the end of the swap and work backwards. At the maturity time t_n, the payoff function for the fixed leg (allowing the floating leg to cancel without penalty) is $\min\{r_n - c, 0\}$. The present value of this payoff at time t_{n-1} is a (short) put option premium, labeled by $-\zeta_{n-1}$. As mentioned in the text, the payoff at time t_{n-1} will become $\min\{r_{n-1} - c - \zeta_{n-1}, 0\}$. The present value of this payoff at time t_{n-2} becomes a simple put and a compound put. Factoring out the discount factor as the discount bond price and working directly with the forward-adjusted expectations, we can compute the total option premium, ζ_{n-1}, at time t_{n-2} as follows:

$$\begin{aligned}
\zeta_{n-2} &= \overline{E}_{n-2}\left[\min\{r_{n-1} - c - \zeta_{n-1}, \ 0\}\right] \\
&= \int_0^{c+\zeta_{n-1}} (r_{n-1} - c - \zeta_{n-1})\overline{\varphi}(r_{n-1}|r_{n-2})dr \\
&= \int_0^{c+\zeta_{n-1}} (r_{n-1} - c)\overline{\varphi}(r_{n-1}|r_{n-2})dr \\
&\quad - \int_0^{c+\zeta_{n-1}} \int_0^{c+\zeta_n} (r_n - c)\overline{\varphi}(r_n|r_{n-1})dr_n \overline{\varphi}(r_{n-1}|r_{n-2})dr_{n-1}.
\end{aligned}$$

Since the interest rate process is assumed to be Markovian, the density function in the second integral can written as a conditional joint density function of r_n and r_{n-1} given r_{n-2}, i.e., $\overline{\varphi}(r_n|r_{n-1})\overline{\varphi}(r_{n-1}|r_{n-2}) = \overline{\varphi}(r_n, r_{n-1}|r_{n-2})$. As a result, we have the first univariate integral as a simple put option and the second bivariate integral as a compound put option. If the interest rate process follows a Gaussian process such as the OU process, then the first integral can be represented as a normal probability function and the second integral as a bivariate normal probability function.

Continue to work backwards, we can arrive at swap value at current time t as follows:

$$\zeta_t = P(t, t_n)\sum_{i=1}^{n} \int_0^{c+\zeta_1} \cdots \int_0^{c+\zeta_i} (r_i - c)\overline{\varphi}_i \prod_{j=1}^{i} dr_j$$

where $\overline{\varphi}_i = \overline{\varphi}(r_1, \cdots, r_i | r_t)$.

CHAPTER 5
PARAMETER ESTIMATION

Estimation of parameters is important because without good estimates, none of the above models in the previous chapters can work properly. Furthermore, hedges can not be constructed if there is a lack of parameters. More importantly, with parameter estimates, one can investigate interest rate dynamics and forecast future interest rate levels. Therefore, in this chapter, we introduce several estimation techniques having been developed recently. The regression method employs the nice closed density function of the state variable. It is easy and convenient. However, it cannot handle the measurement error in the data. It takes a short term interest rate as the instantaneous rate. In doing so, it also loses the ability of estimating the risk premium.

The maximum likelihood estimation is a powerful method. It handles bond prices directly and therefore it is possible to estimate the risk premium along with other parameters of the interest rate process. The drawback of the method is that the structure of the measurement error needs to be simple (to be shown later) otherwise the estimation is messy.

The generalized method of moments is distribution-free and does not rely on any assumption of the interest rate. Consequently it can be used for more than Vasicek and CIR models. However, the power of the method is weak and different moments selected will change the result.

The state space model with Kalman filter is so far regarded as the best model of estimation. It is powerful because it uses a quasi maximum likelihood. Its measurement error structure is not limited. And its factor values are computed from the Kalman filter. It can handle both linear filter of Vasicek and nonlinear filter of CIR. There are ready packages in statistics software like SAS for users. So it is also easy to use.

5.1 SIMPLE REGRESSION

5.1.1 OU Process

The mean-reverting normal process was first adopted by Vasicek (1977) to model the term structure of pure discount bonds. The instantaneous risk-free rate must satisfy the following stochastic differential equation.

$$dr = \alpha(\mu - r)dt + \sigma dW$$

where α, μ, and σ are constants and $dW(t)$ is a normal random variable with mean 0 and variance dt. From Chapter 2, the conditional density of any future instantaneous rate is a normal distribution with the mean and variance as follows:

$$\begin{cases} E_t[r(s)] = r(t)e^{-\alpha(s-t)} + \mu\left(1 - e^{-\alpha(s-t)}\right) \\ \text{var}_t[r(s)] = \dfrac{\sigma^2\left(1 - e^{-2\alpha(s-t)}\right)}{2\alpha}. \end{cases}$$

With this result, we can write the equation as a discrete autoregressive process of order 1, i.e., AR(1) process:

$$r(s) = r(t)e^{-\alpha(s-t)} + \mu\left(1 - e^{-\alpha(s-t)}\right) + e(s)$$

where the error term ε is normally distributed with mean 0 and variance as described. The AR(1) process allows r to satisfy all three properties of the OU process, i.e., mean, variance, and white noise with normal density. Obtaining this exact form from discretization is essential for simplifying the estimation process of the parameters. Equation (3) can be written as the following regression model:

$$r_t = a + b \cdot r_{t-\Delta t} + e_t$$

where

$$a = \mu(1 - b)$$
$$b = e^{-\alpha\Delta t}.$$

One can use a liquid short term interest rate to run this regression. The slope coefficient is used to compute α and then combined with the intercept to compute μ. The σ parameter can be estimated from the mean square error of the regression. Because the error term is normally distributed, this regression method is consistent with the maximum likelihood method. The major disadvantage with this approach is that there is no risk premium in the equation. This is because that we approximate the "rate" with a "price". Note that the state variable itself carries no risk premium; it is the person who trades the asset who imposes his risk preference on the price. Rate quotes observed in the marketplace are actually "prices" which reflect traders' risk preferences. To overcome this problem, we need to apply a complete maximum likelihood estimation (MLE) process or a generalized method of moments (GMM). A complete MLE is a nonlinear estimation process in which choosing an initial value to guarantee convergence is a difficult task. On the other hand, the generalized method of moments can generate different results when different moments are selected. Nonetheless, although we preserve simplicity in this paper, our results are remarkably close to previous studies that used either MLE or GMM.

A simplifying assumption is made to permit use of equation (4). A short term interest rate such as 3-month Treasury bill is used as a proxy of the instantaneous rate. The problem with such an assumption is not so much the measurement error introduced by using a proxy as it is the loss of the ability to estimate the market price of risk. This assumption is not, however, as bad as it first appeared. Since most theoretical pricing models need to allow some flexibility in parameters for future price adjustments;[17] the risk preference parameter can serve this purpose. Use of the implied value generates good results in prediction.[18]

5.1.2 SR Process

In solving the problem of negative interest rates in the Vasicek model, Cox, Ingersoll, and Ross (1985), propose another desirable mean-reverting process for the instantaneous risk-free rate. This process must satisfy a different stochastic differential equation as follows:

$$dr = \kappa(\theta - r)dt + \upsilon\sqrt{r}dW.$$

Following Feller (1951), Cox, Ingersoll, and Ross (CIR) also demonstrate that this process has a conditional density that is proportional to a non-central chi-square distribution as follows:

$$f(r(s)|r(t)) = ce^{-c(r(s)+\xi)}\left(\frac{r(s)}{\xi}\right)^{d/2} I_d\left(2c\sqrt{\xi r(s)}\right)$$

where

$$d = \frac{2\kappa\theta}{\upsilon^2} - 1$$

[17] This is similar to that people use the Black-Scholes formula to imply volatility instead of using the historical estimate.

[18] Although not reported, separate tests have been performed by assuming fixed risk preference over time. The results are significantly improved when the implied risk preference method is used.

$$c = \frac{2\kappa}{\upsilon^2\left(1-e^{\kappa(s-t)}\right)}$$

$$\xi = e^{-\kappa(s-t)}r(t).$$

The mean and variance of this distribution can be obtained as

$$E[r(s)|r(t)] = r(t)e^{-\kappa(s-t)} + \theta\left(1-e^{-\kappa(s-t)}\right)$$

$$V[r(s)|r(t)] = r(t)\frac{\upsilon^2}{\kappa}\left(e^{-\kappa(s-t)} - e^{-2\kappa(s-t)}\right) + \theta\frac{\upsilon^2}{2\kappa}\left(1-e^{-2\kappa(s-t)}\right)^2.$$

It is clear, then, that the mean of the distribution remains the same as that of the OU process but that the variance has become much more complicated. Specifically, the variance is a function of the state variable and is therefore time-dependent. This complication can also be observed from the differential equations themselves. The drift terms of both equations are the same, but the SR diffusion is a function of the state variable while the OU diffusion term is constant. The more complicated variance structure of the SR process leads to a more complicated estimation procedure such that the AR process is no longer applicable. The regression model that produces the least square results can, however, be modified to produce reasonable estimates. For example, let us adopt the same regression model as the previous section. The error term is no longer *i.i.d.* Rather, it is a function of the state variable; therefore, the method of ordinary least squares does not apply. With the variance structure specified above, however, the regression model can be viewed as a regression model with hetroskedasticity. By adopting the standard technique in Judge *et al.* (1982, p.416), we can run weighted least squares as follows:

1. Run an OLS: $r_t = \beta_0 + \beta_1 r_{t-1} + u_t$
2. The variance of the error term must be:
$$E[u_t^2] = r(t)\frac{\upsilon^2}{\kappa}\left(e^{-\kappa(s-t)} - e^{-2\kappa(s-t)}\right) + \theta\frac{\upsilon^2}{2\kappa}\left(1-e^{-2\kappa(s-t)}\right)^2$$
3. Run $u_t^2 = \alpha_1 + \alpha_2 r_{t-1} + w_t$ where:
$$\alpha_1 = \theta\frac{\upsilon^2}{2\kappa}\left(1-e^{-2\kappa(s-t)}\right)^2 \text{ and } \alpha_2 = \frac{\upsilon^2}{\kappa}\left(e^{-\kappa(s-t)} - e^{-2\kappa(s-t)}\right)$$
4. Solve κ and θ from OLS and use αs to solve for υ.

It is important to note that both αs can give υ. In this instance, we choose the one that generates a more consistent estimate with previous studies.

As with the OU process, the price of risk is not identified because T-bill rates are used as a proxy for the instantaneous risk-free rate. In the fitting of the term structure, we use the variable flexibly to adjust properly for theoretical prices.

5.1.3 Some Results

Monthly three-month T-bill yield series from January 1960 to July 1993 are taken from Citibase. Observations from 1960 to 1987 (28 years) are used for parameter estimation. The Vasicek term structure model (OU process) uses four parameters, α, μ, σ, and q. The first three parameters characterize the OU process while q is the price of risk reflected in bond prices. The CIR model (SR process), also uses four parameters: κ, θ, υ, and λ. The first three parameters are similar to those in the OU process. However, it is $\upsilon\sqrt{r}$ that describes the interest rate volatility, not υ itself.

Since the data series in Citibase is interest rates, we directly applied estimation equations in the previous section. Except for the HL model, the equations estimate only non-preference parameters. The risk preference parameter is not estimated and is kept flexible for better fitting. With the three-month interest rate series used as a proxy for the instantaneous rate, the following table summarizes the results.

	OU	SR
mean-reverting	0.2456(α)	0.2456(κ)
reverting level	0.06477(μ)	0.06477(θ)
volatility	0.028855(σ)	0.149895(υ)

Note: The period used in this estimation is 1960–1987. The data are monthly three-month interest rates from Citibase.

Several studies in the literature have reported similar results. For the mean reversion of the OU and SR processes, the estimates indicate a fairly fast mean reversion, which is typical in short-term interest rates. The half mean life of the interest rate is 1.22 years. Both parameters are significant at the 99-percent confidence level.

5.1.4 Yield Curve Fitting Exercise

In this section, we perform out-of-sample term structure fitting of the 10-year Treasury bond prices. We conduct the test by using 10-year Treasury bonds for the period January 1988 through July 1993. Since the data from Citibase are 10-year Treasury yields, they must be converted to prices. Given that yield curves are functions of the coupon rate, which is not provided by the database, we assume that the coupon rates are equal to the yields. In other words, we assume that Citibase compiled the interest rate series from par bonds.

The test is performed by incorporating the time-varying risk premia constantly documented in the term structure literature. An appropriate bond pricing model is necessary to estimate the risk premium terms. Depending on the pricing models used in each process, we adopt various methods to compute the theoretical value of the 10-year Treasury bonds. Parameters estimated in the previous section are used for various models.

We allow the risk-preference parameter to be implied by each month's bond price and then use it to price the bond in the next month. Such a procedure has been widely used in the Black–Scholes model for pricing equity options to incorporate the time-varying volatility. We generate 66 implied risk parameters from January 1988 to June 1993 for each process. The following table reports the summary statistics for these parameters.

	OU(q)	SR(λ)
Minimum	–.4190	–0.1719
Maximum	–.1386	–0.0755
Mean	–.2718	–0.1290
Standard Deviation	0.0670	0.0258

Note: These sample statistics of the risk preference parameter for each process are calculated from 66 (1/88 to 6/93) parameter values implied by the model. Other parameters are taken from Table 1.

The implied risk preference term is then used as an input variable in pricing the following month's T-bond value. The theoretical value generated by the model is subtracted from the actual observed market price. This procedure is repeated 66 times, i.e., from February 1988 to July 1993. The following table provides the mean square errors.

	OU	SR
SSE	112.15	107.73
MSE	1.70	1.61

Note: The period used in this estimation is February 1988 to July 1993 (N = 66). All data except the risk preference parameter are taken from Table 1. The risk-preference parameter is implied from the bond price of each month.

5.2 MAXIMUM LIKELIHOOD ESTIMATION

In the above section, the estimation of the Vasicek model is consistent with the maximum likelihood estimation since the error term is normally distributed. Note that the normally distributed error term comes from the assumption that the interest rate is normal, which is different from the regression models we learn in introductory econometrics in which error terms are assumed normal for convenience. For the CIR model, the results will not be consistent with the maximum likelihood. We handle the conditional variance problem with a simple heteroskedesticity structure which does not fully explain the conditional variance. In the above regression models, there is no measurement error. Also, both models estimated above cannot provide the estimate of the risk premium. Next, we show how a complete maximum likelihood estimation procedure is implemented.

5.2.1 Methodology

We use the Vasicek model in this subsection for demonstration while estimation results in the following subsections are based upon CIR's model. It should be noted that the two models are similar in terms of estimation. The Vasicek model is obtained from the previous chapter as:

$$P(t,T) = e^{-r(t)F(t,T)-G(t,T)}.$$

Since there is measurement error in the data, the observed bond price does not match this formula. So we can write the real price as the formula price plus an additional error term:

$$\ln \tilde{P} = \ln P(t,T) + u$$

and this error is assumed normal. The maximum likelihood estimation can be established because the likelihood function can be derived. Brown and Dybvig (1986) find parameter estimates using this formulation. They use cross sectional bond data to minimized the errors and find parameter values. Note that if the model is CIR, then the interest rate is not normally distributed. But here the estimation does not use that distribution information. In the following, we introduce the maximum likelihood that use both the state variable distribution as well as the measurement error distribution.

The joint distribution of the state variables over time for a one-factor model can be written as:

$$f(r_t, r_{t+1}, \cdots, r_T) = \left[f(r_T | r_{T-1}) \cdots f(r_{t+1} | r_t) f(r_t)\right].$$

This follows from the Markov property. In the Vasicek case, each conditional density is a normal distribution (described before). The unconditional distribution $f(r_t)$ is usually replaced by the current interest rate value. Note that r is an instantaneous rate which is not directly observable from the market. Consequently, we need to replace this distribution with a distribution that can be observed. Since the Vasicek model is in closed form, we can do it. Write the pure discount bond model of Vasicek as follows:

$$\ln P(t, t + \Delta t) = -G(t, t + \Delta t) - r(t) F(t, t + \Delta t).$$

Reversing the formula and writing the interest rate in terms of the bond price, we get the following:

$$r(t) = -\frac{\ln P(t, t + \Delta t) + G(t, t + \Delta t)}{F(t, t + \Delta t)}.$$

Since the observations come in equal intervals, we should for notation simplicity drop the argument, $(t, t + \Delta t)$. The Jacobian is:

$$\left|\frac{dr}{d \ln P}\right| = \frac{1}{F}.$$

Substituting back to the likelihood function, we get:

$$\begin{aligned} f(r_t, r_{t+1}, \cdots, r_{t+n}) &= f(r_n | r_{n-1}) \cdots f(r_{t+1} | r_t) f(r_t) \\ &= f(\ln P_n | \ln P_{n-1}) \cdots f(\ln P_{t+1} | \ln P_t) f(\ln P_t) F^{-n}. \end{aligned}$$

The log likelihood function is:

$$\ln L = \Sigma_{j=1}^{n} \ln f(\ln P_{t+j} | \ln P_{t+j-1}) - n \ln F.$$

Taking partial derivatives with respect to parameters, we get a set of 4 simultaneous equations. Solving these equations, we get maximum likelihood estimators for the four parameters. Since it allows no measurement error, only one bond series is enough for the estimation.

Measurement errors are sometimes considered by empiricists for a number of reasons such as discrete tick size and bid ask spread. To include measurement errors in the log likelihood function, we need to increase the number of bonds in the estimation to avoid technical problems. Usually, any additional measurement error requires an additional bond. In Chen and Scott (1993), they use 4 bond price series. When they estimate the one-factor model described above, they assume 3 measurement errors. Similarly, the two-factor model they estimated has 2 measurement errors and the three-factor model has only 1 measurement error.

To exemplify, if one includes 1 measurement error in the model as follows

$$\begin{aligned}\ln \tilde{P}_1 &= \ln P_1(t, T) \\ \ln \tilde{P}_2 &= \ln P_2(t, T) + u\end{aligned}$$

then the above log likelihood function will change to

$$\begin{aligned}\ln L = &\Sigma_{j=1}^{n} \ln f(\ln P_{1,t+j} | \ln P_{1,t+j-1}) - n \ln F_1 \\ &+ \Sigma_{j=1}^{n} \ln f(\ln P_{2,t+j} | \ln P_{2,t+j-1}) - n \ln F_2 - \tfrac{1}{2} \ln 2\pi - \ln \sigma - \frac{u^2}{2\sigma^2}\end{aligned}$$

where u is function of both bond prices. Here we assume a normally distributed error term u with mean 0 and standard deviation σ. Taking the derivatives with respect to parameters and solving the system of nonlinear simultaneous equations will lead to parameter estimators.

For more than one factor, the procedure is exactly the same. If one wants the same measurement error for each additional factor, then the number of bonds would have to increase. Chen and Scott (1993) show a very detailed discussion on this estimation, and interested readers are referred to their paper. In the next section, we will simply report their results on the estimation.

5.2.2 Some Results

The multi-factor Vasicek model has not been estimated yet. Chen and Scott (1993) use the square root process of the CIR model to estimate one, two and three-factor models. The SR process is more complicated because the distribution of the state variables is joint non-central chi-square. The estimator of the parameters are therefore more difficult to compute. Nonetheless, the MLE procedure is exactly the same as the normally distributed OU process.

The data used in the estimation are monthly bond series from 1960 to 1987 with total of 326 months. Bond rates of 3 months, 6 months, 5 years, and the longest maturity available are taken. The parameters for one-, two-, and three-factor models are given in the following table.

One-factor model			
	Variable 1		
κ	0.4697		
θ	0.0618		
υ	0.0825		
λ	–0.0454		
Two-factor model			
	Variable 1	Variable 2	
κ	0.7660	0.0008991	
θ	0.0321	0.0212	
υ	0.1312	0.0531	
λ	–0.1186	–0.0415	
Three-factor model			
	Variable 1	Variable 2	Variable 3
κ	1.6331	0.005064	0.006237
θ	0.0324	0.0108	0.0091
υ	0.1373	0.0755	0.1842
λ	–0.0317	–0.1530	–0.1373

Note: These estimates are obtained from a monthly data set from 1906–1987 with sample size N=326.

It is seen that for the one-factor model, the mean reverting speed is only 0.4697 but with more factors, the first factor has faster speed. The second or both the second and the third have very slow mean reversion which behaves very much like a random work. The reverting level is 6.18% for the one-factor model, 5.33% (3.21%+2.12%) for the two-factor model, and 4.41% (3.24%+1.08+0.91) for the three-factor model. The volatility parameter υ behaves similarly to the speed parameter. The first volatility increases when more factors are considered. However, in the three-factor model, the third factor accounts for the most short rate volatility.

5.2.3 Yield Curve Fitting

The fitting of the yield curves of the selected dates for the one-, two-, and the three-factor models is summarized in the following table. Chen and Scott choose all bonds of the first Thursdays of June and December from 1980 to 1988. Using the parameter values in the previous table, the root mean square errors are computed by comparing the observed mid bond prices with the model prices.

Total of 2,341 bonds	Absolute Pricing Errors	Percentage Pricing Errors	Yield-to-Maturity (bps)
One-factor model	$3.89	3.89%	95.16
Two-factor model	1.37	1.35	51.09
Three-factor model	0.91	0.97	40.05

It is clear that more factors can fit better the yield curves. It is also seen that from one to two factors, the errors reduce dramatically (about halved) while from two to three factors, the errors reduce only marginally. Selected plots can be found in Chen and Scott's original article.

5.3 GENERALIZED METHOD OF MOMENTS

5.3.1 Methodology

GMM is a very useful tool when distributions are unknown. Matching sample and theoretical moments can give estimates to parameters. In its simplest form, the method of moments chooses the number of moments to be the same as the number of parameters. In general, the selection criterion is from low to high orders of moments. In estimating term structure models, however, it is not so simple. We would like to incorporate as many bonds as possible in the estimation. Gibbson and Ramaswamy (1993) use this technique to estimate the one factor CIR model. Their procedure goes as follows. They define the holding period return (1 plus rate of return) between now (time t) and the next period (time $t+1$) as:

$$R(t,t+1) = \frac{P(t+1,T)}{P(t,T)}$$

where $P(t,T)$ is the price at time t of a pure discount bond that pays $1 at time T. If there is a pricing model for the discount bond price, then the holding period return defined above will have a closed form. Then, we can calculate its unconditional mean as:

$$E[R(t,t+1) = E\left[\frac{P(t+1,T)}{P(t,T)}\right]$$

for any given bond. Similarly, the unconditional covariance between any two different-maturity bonds can also be found as:

$$E[R_a(t_1,t_2)R_b(t_3,t_4)] = E\left[\frac{P(t_1,T_a)}{P(t_2,T_a)} \cdot \frac{P(t_3,T_b)}{P(t_4,T_b)}\right]$$

With a collection of different bonds over time, we can match sample moments (means and covariances) with the two formulas given above and estimate the parameters in the model.

Take the Vasicek model as an example. We know from Chapter 2 that the Vasicek model has the following closed form solution:

$$P(t,T) = e^{-r(t)F(t,T)-G(t,T)}$$

where

$$F(t,T) = \frac{1-e^{-\alpha(T-t)}}{\alpha}$$
$$G(t,T) = \left(\mu - \frac{q\sigma}{\alpha} - \frac{\sigma^2}{2\alpha^2}\right)[T-t-F(t,T)] + \frac{\sigma^2 F^2(t,T)}{4\alpha}.$$

Therefore, the conditional mean of the holding period return on time t is given as:

$$\begin{aligned} E_t\left[\frac{P(t+1,T)}{P(t,T)}\right] &= \frac{1}{P(t,T)} E_t[P(t+1,T)] \\ &= \frac{e^{-G(t+1,T)}}{P(t,T)} e^{-E_t[r(t+1)]F(t+1,T)+F(t+1,T)^2 V_t[r(t+1)]} \end{aligned}$$

The second line on the right hand side is the result of the moment generating function. Also from Chapter 2, we know that the conditional mean and variance are (on the original probability space):[19]

$$\begin{cases} E_t[r(s)] = r(t)e^{-\alpha(s-t)} + \mu\left(1-e^{-\alpha(s-t)}\right) \\ K_t[r(u),r(s)] = \text{cov}_t[r(u),r(s)] = \frac{\sigma^2}{2\alpha} e^{-\alpha(s+u-2t)}\left(e^{2\alpha(u-t)}-1\right) \quad \text{for } u < s. \\ V_t[r(s)] = \text{cov}_t[r(s),r(s)] = \frac{\sigma^2\left(1-e^{-2\alpha(s-t)}\right)}{2\alpha} \end{cases}$$

[19] If the expectation of the bond price is taken in the risk-neutralized space, then it is a futures price.

Substituting this result back to the conditional holding period return, we shall get:

$$E_t\left[\frac{P(t+1,T)}{P(t,T)}\right]=$$
$$e^{-r(t)\left[F(t+1,T)\left(e^{-\alpha\Delta t}\right)-F(t,T)\right]-\left[G(t+1,T)-G(t,T)+F(t+1,T)\mu\left(1-e^{-\alpha\Delta t}\right)+F(t+1,T)^2\sigma^2\left(1-e^{-2\alpha\Delta t}\right)/(2\alpha)\right]}$$

where Δt is the time period between t and t+1. For notation convenience, we rewrite the above equation as:

$$E_t[R(t,t+1)]=e^{-r(t)X-Y}.$$

where

$$X=F(t+1,T)\left(e^{-\alpha\Delta t}\right)-F(t,T)$$
$$Y=G(t+1,T)-G(t,T)+F(t+1,T)\mu\left(1-e^{-\alpha\Delta t}\right)+F(t+1,T)^2\sigma^2\left(1-e^{-2\alpha\Delta t}\right)/(2\alpha)$$

The unconditional needs to integrate this conditional expected value over the unconditional density of $r(t)$. The unconditional density is usually the steady state distribution of the stochastic process. In other words, by letting s approach infinity, we obtain the unconditional density for the state variable as a normal density with the mean and variance as:

$$\begin{cases} E[r(t)]=\mu \\ V[r(t)]=\frac{\sigma^2}{2\alpha} \end{cases}.$$

As a result, the unconditional mean can be solved as:

$$E[R(t,t+1)]=e^{-\left(Y+\alpha\sigma^2X^2+\mu X\right)}.$$

A similar procedure is taken for solving the covariance. We first calculate the conditional covariance:

$$E_t\left[\frac{P(t_1,T_a)}{P(t_2,T_a)}\cdot\frac{P(t_3,T_b)}{P(t_4,T_b)}\right]=$$
$$E_t\left[e^{-r(t_1)F(t_1,T_a)-G(t_1,T_a)+r(t_2)F(t_2,T_a)+G(t_2,T_a)-r(t_3)F(t_3,T_b)-G(t_3,T_b)+r(t_4)F(t_4,T_b)-G(t_4,T_b)}\right]=$$
$$e^{\Lambda-r(t_1)F(t_1,T_a)+r(t_2)F(t_2,T_a)-r(t_3)F(t_3,T_b)+r(t_4)F(t_4,T_b)}$$

where

$$\Lambda = -G(t_1, T_a) + G(t_2, T_a) - G(t_3, T_b) - G(t_4, T_b)$$

The random variable $-r(t_1)F(t_1, T_a) + r(t_2)F(t_2, T_a) - r(t_3)F(t_3, T_b) + r(t_4)F(t_4, T_b)$ is normal with mean and variance as:

$$M = -r(t)\left[e^{-\alpha(t_1-t)}F(t_1, T_a) + e^{-\alpha(t_3-t)}F(t_3, T_b) - e^{-\alpha(t_2-t)}F(t_2, T_a) - e^{-\alpha(t_4-t)}F(t_4, T_b)\right] - \mu\begin{bmatrix}\left(1-e^{-\alpha(t_1-t)}\right)F(t_1, T_a) + \left(1-e^{-\alpha(t_3-t)}\right)F(t_3, T_b) - \\ \left(1-e^{-\alpha(t_2-t)}\right)F(t_2, T_a) - \left(1-e^{-\alpha(t_4-t)}\right)F(t_4, T_b)\end{bmatrix}$$

$$V = \sum_{i=1}^{4}\sum_{j=1}^{4} (-1)^{i+1} F(t_i, T_i)F(t_j, T_j)K_t[r(t_i), r(t_j)]$$

where $T_1 = T_2 = T_a$ and $T_3 = T_4 = T_b$. For notation convenience, we write the conditional covariance as:

$$E_t[R(t_1, t_2)R(t_3, t_4)] = e^{\Lambda - M + V/2} = e^{-r(t)A - B}$$

where

$$A = e^{-\alpha(t_1-t)}F(t_1, T_a) + e^{-\alpha(t_3-t)}F(t_3, T_b) - e^{-\alpha(t_2-t)}F(t_2, T_a) - e^{-\alpha(t_4-t)}F(t_4, T_b)$$

$$B = -\Lambda - \frac{V}{2} + \mu\begin{bmatrix}\left(1-e^{-\alpha(t_1-t)}\right)F(t_1, T_a) + \left(1-e^{-\alpha(t_3-t)}\right)F(t_3, T_b) - \\ \left(1-e^{-\alpha(t_2-t)}\right)F(t_2, T_a) - \left(1-e^{-\alpha(t_4-t)}\right)F(t_4, T_b)\end{bmatrix}$$

Although A and B are more complex than X and Y, the exponential form remains the same, which implies that the unconditional expectation of the product of two returns should take the same form as the unconditional expected return. We write the final result as:

$$E[R(t_1, t_2)R(t_3, t_4)] = e^{-\left(B + \alpha\sigma^2 A^2 + \mu A\right)}$$

We have finished deriving the theoretical moments of the Vasicek model. By matching these moments with sample moments from data, we can estimate the parameter values with proper iterations.

It should be noted that the theoretical moments of the CIR model are significantly more difficult to derive, although the solution procedure is identical. This is because the steady state distribution is a Gamma function which is more complicated than the normal function.

With multiple factors, it is noted that since factors are assumed independent to one another, the unconditional expectations can be performed for each individual factor. For example, assume a two factor model:

$$r = y_1 + y_2$$

where each y follows an OU process. Then the unconditional expectations become:

$$E[R(t,t+1)] = e^{-\left(Y_1+\alpha_1\sigma_1^2X_1^2+\mu_1X_1\right)-\left(Y_2+\alpha_2\sigma_2^2X_2^2+\mu_2X_2\right)}$$

for the mean and:

$$E[R(t_1,t_2)R(t_3,t_4)] = e^{-\left(B_1+\alpha_1\sigma_1^2A_1^2+\mu_1A_1\right)-\left(B_2+\alpha_2\sigma_2^2A_2^2+\mu_2A_2\right)}$$

for the covariance.

5.3.2 Some Results

Gibbons and Ramaswamy (1993) estimated the one factor CIR model using GMM. The data used in their study are monthly holding period returns on 1-, 3-, 6-, and 12-month U.S. Treasury bills. This result can be compared with the estimates in the previous and next sections.

One-Factor model	
κ	12.43
θ	0.0154
υ	0.49
λ	–6.08

Note: Data used in this study are monthly holding period returns on 1-, 3-, 6-, and 12-month US Treasury bills from 1964 through 1989.

The GMM estimates essentially finds the parameter values that will set these sample moments as close to their theoretical values as is possible. In general, GMM is not an appropriate method for parameter estimation of the models discussed in this book. It is good for other purposes such as relationships among different rates. The models introduced here all have closed form density functions and MLEs are much more powerful than GMM estimators. In the next section, it is shown that the state space model is a very good choice for our purpose.

5.4 STATE SPACE MODEL WITH KALMAN FILTERING

The problem of the maximum likelihood method is that due to a mapping problem the number of bonds needs to equal the number of factors plus the number measurement errors. As Chen and Scott indicates, with only 4 bonds throughout for one-, two-, and three-factor models, they face a trade-off between the factors and measurement errors. Furthermore, the calculation of the factor values seems quite arbitrary. In their paper, they find the factor values by matching two bonds (long and short) exactly.

Although the generalized method of moments is free from the above problems, the selection of the moments is quite arbitrary and the power of the method is less than that of the MLE method. Recently Chen and Scott (1995) improve the maximum likelihood method by fitting the CIR valuation equation into a state space model by Engle and Watson (1981). With this model, there is no limitation on the structure of the measurement error while the MLE remains valid. The factor values are then retrieved by the Kalman filter.

5.4.1 Methodology

To be general, we discuss the multi-factor model which is consistent with previous discussions. Each state variable, y_j, is assumed to follow a risk-neutralized mean reverting normal (or Ornstein–Uhlenbeck) process:

$$dy_j = (\alpha_j \mu_j - \alpha_j y_j - \sigma_j q_j)dt + \sigma_j d\hat{W}_j.$$

The instantaneous interest rate in this model is always non-negative. The time t price for a discount bond that pays \$1 at time s is:

$$P(t,s) = e^{-\sum_{j=1}^{J} F_j(t,s)y_j - G_j(t,s)}.$$

The (continuously compounded) yield for a discount bond is computed as follows:

$$R(t,s) = \frac{-\ln P(t,s)}{s-t} = \frac{-1}{s-t}\sum_{j=1}^{J} G_j(t,s) - F_j(t,s)y_j.$$

The fixed parameters for this model can be estimated from a time series of bond rates. This equation shows that bond returns are linear functions of state variables.

The solutions for bond rates can be used to estimate the fixed parameters in the model. As noted earlier, each state variable has a conditional distribution that is normal. The expectation of y_j at time s, conditional on knowing y_j at time t, and the conditional variance are equal to:

$$\mu_j\left(1 - e^{-\alpha_j(s-t)}\right) + e^{-\alpha_j(s-t)}y_{j,t}$$

and

$$\frac{\sigma_j^2\left(1 - e^{-2\alpha_j(s-t)}\right)}{2\alpha_j}.$$

If we consider the bond rates and the state variables taken at discrete points in time, we have the following linear model:

$$\begin{aligned} y_{j,t} &= \theta_j\left(1 - e^{-\alpha_j \Delta t}\right) + e^{-\alpha_j \Delta t} y_{j,t-1} + e_{j,t} \\ &= a_j + \phi_j y_{j,t-1} + e_{j,t} \\ R(t,s_m) &= \sum_{j=1}^{J} -\frac{G_j(t,s_m)}{s_m - t} + \sum_{j=1}^{J} \frac{F_j(t,s_m)}{s_m - t} y_{j,t} \end{aligned}$$

where Δt is the size of the time interval between the discrete observations. The innovation for each state variable, e_{jt}, has a conditional mean of zero and a conditional variance equal to the conditional variance for the state variable given in (4). If we add a measurement error to each observed bond rate, the model in (5) becomes a state space model. A Kalman filter can be used to compute estimates of the unobservable state variables at each time period, and quasi maximum likelihood can be used to estimate the fixed parameters.[20]

[20] See Chen, R. and L. Scott, "Multi-Factor Cox-Ingersoll-Ross Models of the Term Structure of Interest Rates: Estimates and Tests Using a Kalman Filter," Working Paper, Rutgers University and University of Georgia, 1993 and Chen, R. and L. Scott, "Maximum Likelihood Estimation for a Multi-factor Equilibrium Model of the Term Structure of Interest Rates," *Journal of Fixed Income*, December, 1993.

5.4.2 Some Results

Chen and Scott (1995) have used this state space model to estimate one-, two-, and three-factor models from a monthly data set of bond rates for the period 1960–87. In a factor analysis study, Litterman and Scheinkman (1991) have found that three factors explain most of the variation in bond returns. In following table, we report the Chen and Scott results for the one-, two-, and three-factor CIR models.

The factor estimates that are computed from the Kalman filter are related to several familiar aspects of the term structure. If the factors are normally distributed (Vasicek), then the standard linear Kalman filter can be readily used. Statistical packages like SAS contains standard routines. However, in the CIR model, factors are not normally distributed. They are confined to be non-negative. As a result, the filter used in this case is a nonlinear filter. Chen and Scott argue that for the CIR model, this problem can be easily solved by replacing negative factor values with 0. They prove that this procedure produces efficient estimates.

One-Factor model

	Variable 1
κ	0.07223
θ	0.03739
υ	0.07540
λ	–0.07892

Two-Factor Model

	Variable 1	Variable 2
κ	0.64060	0.01703
θ	0.03082	0.00003
υ	0.12790	0.05541
λ	–0.17410	–0.04073

Three-Factor Model

	Variable 1	Variable 2	Variable 3
κ	1.36820	0.08436	0.00844
θ	0.02979	0.00065	0.00072
υ	0.1231	0.1355	0.04883
λ	–0.32290	–0.04427	–0.05831

Note: These estimates are obtained from a monthly data set from 1906–1987 with sample size N=326.

Note that factors in the model are neither observable nor interpretable. However, Chen and Scott run a series of correlation checks and identify comovement between factors and interest rate variables. They argue that the first factor in the three-factor models is highly correlated with the slope of the term structure: the factor moves up and down with the difference between the short term rate and the long term rate.[21] The first factor has strong mean reversion which is also a characteristic of the slope of the term structure. The second factor in the three-factor model is highly correlated with the 5 year bond rate.[22] The third factor in the three-factor model displays less variability; in the three-factor model, the second and third factors together determine medium and long term bond rates as well as the volatility of rates. These factors present weak mean reversion, which is consistent with the random walk characteristics of these bond rates.

5.4.3 Yield Curve Fitting

The fitting of the yield curves of the selected dates for the one-, two-, and the three-factor models is summarized in the following table. Chen and Scott choose all bonds of the first Thursdays of June and December from 1980 to 1988. Using the parameter values in the previous table, the root mean square errors are computed by comparing the observed mid bond prices with the model prices.

Total of 2,341 bonds	Absolute Pricing Errors	Percentage Pricing Errors	Yield-to-Maturity (bps)
One-factor model	\$3.00	3.12%	86.67
Two-factor model	1.34	1.35	47.45
Three-factor model	0.87	0.89	38.32

Comparing this table with the table of the MLE in the previous section, we see little difference. However, as Chen and Scott (1993) noted, in the MLE method, the log likelihood function is less for the three-factor model than the two-factor model, meaning the three-factor is worse in terms of fitting the joint distribution produced by factors and measurement errors, although the yield curve fitting is better. Here, since there is no trade-off between the measurement errors and the state variables, the three-factor model simply dominates the two-factor model.

5.5 SUMMARY

In this chapter, we summarize the current methodologies used in estimating

[21] This variable in the graphs, the short rate minus the long rate, is the negative of the slope of the term structure.
[22] Although it is not shown in the graph, the 5 year bond rate and the long term rate tend to move together.

parameters of the term structure models. It should be noted that only fixed-parameter models can be estimated econometrically. Time-dependent-parameter models, on the other hand, can hardly be estimated with sound econometric methods. Only after parameters being estimated reliably, we can then price and hedge interest rate risks with accuracy. We have reviewed the regression method, the MLE method, the GMM method, and the state space model. The current research seems to indicate that the state space model with Kalman filtering is the most robust method to use. However, we can also find that simple method such as the regression method can produce similar results. From pricing's perspective, I think we need to worry more about if the fundamental term structure model is correct than if parameters are estimated with 100% accuracy. There have been studies recently showing long memory, significant conditional heteroskdesticity, and cointegration. How these characteristics are incorporated in future term structure models is perhaps a more important question to answer.

CHAPTER 6
HEDGING INTEREST RATE RISKS

6.1 INTRODUCTION

People normally find it hard to believe that oversimplified models like the ones introduced in this book can be really used in pricing a security. We know that these models may be wrong in the distribution assumption, wrong in volatility structure, wrong in Markov property, wrong in fixed parameters, or maybe wrong in the basic the term structure theory. Then why should we believe that prices calculated by these models should reflect the "right" price? The answer is no, we don't! These models should not be used to calculate actual prices. Actually, no theoretical models should be used in calculating prices. If models are not used in pricing, then what are they used for? The answer is: hedging.

In finance, we do not question market prices because they are determined by the smartest people, traders, who make life-or-death decisions everyday. Models should be able to match these market prices. Parameters in the models are set so that they generate market prices. Once the market prices are matched, we look at the models and ask what hedge ratios these models tell us.

If a model, although simple, can capture the most important characteristics of the underlying risks, hedges suggested by the model will be robust and reliable. For example, even though the Black–Scholes model presents a lot of pricing biases like volatility smile and the volatility curve, it has been used quite successfully. If a model is simple and has a closed-form solution, then the hedge ratio is easy to calculate. If a model does not have a closed-form solution, then the hedge ratio needs to be computed numerically. Sometimes this is slow and may not be good enough for traders. That is why closed-form solutions are important; not because they are elegant, but because they are fast.

Single-factor models are believed impossible to fit the market data, no matter how you change the combination of the parameters. As a result, single-factor models are useful only in exposition and communication. There is no real use of the single-factor models. In the interest rate markets, a good model needs to do at least two jobs: to fit the yield curve and to fit the volatility curve. In plain language, a good term structure model should be able to fit all bond prices of different maturities and it should give a good option formula to fit all option prices of different maturities. If single-factor models cannot do the work, then what models should we use?

There are two approaches, one is to add some flexibility in the model so that fitting curves becomes no problem. Ho and Lee (1986), Hull and White (1990) and Heath, Jarrow, and Morton (1992) follow this approach. It is called time-dependent parameter models. Another approach is to allow more than one state variable for the term structure. This approach is used by Langtieg (1980), Chen and Scott (1992) and Longstaff and Schwartz (1992). This should create more flexibility and should improve the fitting. The pros and cons of each approach are provided in the last chapter in which we cite a recently developed model.

No matter which approach is adopted in modeling the term structure, we need to understand its hedges. Since all models are developed in continuous time, the hedging implications are also in continuous time; or over short intervals. Again, for exposition purposes, we shall start our discussion with a single-factor model. Once the idea is brought across, we shall give examples in both multi-factor models and time-dependent parameter models

6.2 DELTA–GAMMA HEDGING

Since options are not linear functions of the underlying asset, hedging option-like contracts becomes complex. Delta hedging, which produces a linear hedge cannot be sufficient if there is a significant swing in the spot market. Gamma hedging therefore becomes necessary.

Given an option-like contract, the delta hedging is to trade its underlying asset as frequently as possible to guarantee that the option-like contract does not lose value when the spot market fluctuates. To exemplify, we form a portfolio of the option-like contract and its underlying asset and the portfolio value is immune to the price change of the underlying asset. Formally, set the portfolio as $V = C + h \times S$ and

$$dV = dC + h \times dS = 0.$$

Therefore, the hedge ratio is $-dC/dS$. Note that dV can remain 0 only when h is adjusted to market changes frequently enough. The following graph tells how effective the delta hedging is:

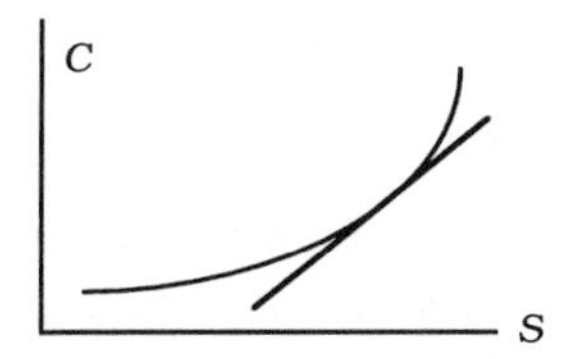

Often, the portfolio to be hedged contains many complex securities and it is not possible to find the direct first order derivative as described above. Under this circumstance, we need to find the underlying risk factor we want our portfolio to immune to and find the first order derivative with respect to this risk factor. The hedging security is also measured with respect to this risk factor. Then, the ratio of the two first order derivatives provides the hedge ratio. Let us look at an example below.

[EXAMPLE 1]
A 5 year, 6% coupon bond is sold exactly at $96.79 which is correctly priced by the Vasicek model (coupons paid annually). How can use a one–year Treasure bill to hedge this bond?

It is clear that the coupon bond is not a direct function of the one–year Treasury bill. However, the coupon bond is a function of the instantaneous rate. Assume a one–factor–model of Vasicek (using the parameter values in Example 2 of Chapter 2), we can find the derivative of the coupon bond with respect to the instantaneous rate as:

$$\begin{aligned} Q(t,\underline{T}) &= 6[P(0,1)+P(0,2)+P(0,3)]+100P(0.3) \\ \frac{dQ}{dr} &= 6\left[\frac{dP(0,1)}{dr}+\frac{dP(0,2)}{dr}+\frac{dP(0,3)}{dr}\right]+100\frac{dP(0,3)}{dr} \\ &= -6[P(0,1)F(0,1)+P(0,2)F(0,2)+P(0,3)F(0,3)]-100P(0.3)F(0,3) \\ &= -195.6665. \end{aligned}$$

The one–year T bill has the following derivative:

$$\frac{dP(0,1)}{dr} = -P(0,1)F(0,1) = -0.8317$$

In order to hedge the coupon bond with the Treasury bill, for every $100 bond, we need about $235 of Treasury bills.

$$\frac{dQ}{dr} \Big/ \frac{dP(0,1)}{dr} = \frac{-195.6665}{-0.8317} = \$235.2742$$

If the market has a big swing, then h cannot be adjusted in time to maintain a hedged position and the portfolio will lose value. As a result, when traders sense some unusual sentiment in the marketplace, they usually perform Gamma hedging.

Since Gamma hedging is to hedge big swing, called curvature or convexity–concavity, we need a second contract that has curvature, in other words, another option-like contract. To form a portfolio of $V = C + h_1 S + h_2 X$. To maintain Delta-neutral, we need $dV = 0$. To also maintain Gamma-neutral, we need:

$$\frac{d^2V}{dS^2} = 0 = \frac{d^2C}{dS^2} + 0 + h_2 \frac{d^2X}{dS^2}$$

$$h_2 = -\frac{\gamma_c}{\gamma_x}.$$

Therefore,

$$\frac{dV}{dS} = 0 = \frac{dC}{dS} + h_1 + h_2 \frac{dX}{dS}$$

$$h_1 = -\frac{dC}{dS} - h_2 \frac{dX}{dS} = -\delta_c - \delta_x \frac{\gamma_c}{\gamma_x}.$$

In this case, the portfolio is immune not only to δ risk but also γ risk. When there are large changes in the stock market, the option position is still insured.

6.3 SINGLE-FACTOR HEDGING

Since we need to come up with exact hedges, for simplicity, we use the Vasicek model for example. The single-factor Vasicek model gives the option formulas as follows:

$$C(t) = P(t,T)\, N(d) - P(t,T_c)\, K\, N(d - \sqrt{V_p})$$

where

$$d = \frac{1}{\sqrt{V_p}} \left(\ln \frac{P(t,T)}{P(t,T_c)\, K} + \frac{V_p}{2} \right)$$

$$V_p = \mathrm{var}[\ln P(T_c,T)] = F(T_c,T)^2 \sigma^2 \left(1 - e^{-2\alpha(T_c - t)}\right) / 2\alpha.$$

If we take a position in an option contract, then to hedge it, we need to use a traded security. It is very natural to use its underlying asset as the hedging security. To maintain a Delta-neutral hedge, we simply set the portfolio in the following way so that it is insensitive to the interest rate change:

$$V = C - h \cdot P$$
$$\frac{dV}{dr} = \frac{dC}{dr} - h \cdot \frac{dP}{dr} = 0.$$

Therefore, the hedge ratio can be set to:

$$h = \frac{dC}{dP}.$$

By the above closed-form solution, we get:

$$h = N(d).$$

This result is similar to the Black and Scholes hedge where the hedge ratio is also equal to the first probability. Of course this is because in the Vasicek model, the bond is log-normally distributed.

In the CIR model, the result is quite different. In the CIR model, the option cannot be written as a function of the discount bond. Recall the CIR option model in Chapter 2:

$$C(t) = P(t,T)\chi^2(r^*) + P(t,T_c)K\chi^2(r^*)$$

where

$$\chi_j^2(r^*) = n.c.\chi^2\left(2(\eta + B(T_j))r^*; \frac{4\kappa\theta}{\sigma^2}, \frac{2\eta^2 e^{-(\kappa+\lambda)T_f} r}{\eta + B(T_j)}\right),$$
$$\eta(t,T_f) = \frac{2(\kappa+\lambda)}{\sigma^2(1-e^{-(\kappa+\lambda)(T_f-t)})}.$$

In this case, it is impossible for the call to take a partial derivative of the bond price. And we need to take two separate derivatives with respect to the interest rate. Explicitly, we need to compute:

$$\frac{dC}{dP} = \frac{dC}{dr} \Big/ \frac{dP}{dr}.$$

Since both the call and bond formulas are closed-form equations, this hedge can be computed analytically.

6.4 MULTI-FACTOR HEDGING

In a multi-factor setting, the derivatives become a little complex. Take the Vasicek model for example, although the hedge remains $N(d)$, the content is different. The volatility in d is now:

$$
\begin{aligned}
V_p &= \mathrm{var}[\ln P(T_c, T)] \\
&= \mathrm{var}\left[-y_1 F_1(T_c, T) - y_2 F_2(T_c, T)\right] \\
&= F_1(T_c, T)^2 \frac{\sigma_1^2\left(1 - e^{-2\alpha_1(T_c - t)}\right)}{2\alpha_1} + F_2(T_c, T)^2 \frac{\sigma_2^2\left(1 - e^{-2\alpha_2(T_c - t)}\right)}{2\alpha_2}
\end{aligned}
$$

which is more complex.

In the CIR model, the hedge needs two different securities, because it can no longer take derivative with respect to r. Neither the bond formula nor the option formula is a function of r. They are functions of the factors, y_1 and y_2. Look at the portfolio:

$$
V = C - h_1 \cdot P_1 - h_2 \cdot P_2
$$

$$
\begin{cases}
\dfrac{dV}{dy_1} = \dfrac{dC}{dy_1} - h_1 \cdot \dfrac{dP_1}{dy_1} - h_2 \cdot \dfrac{dP_2}{dy_1} = 0 \\
\dfrac{dV}{dy_2} = \dfrac{dC}{dy_2} - h_1 \cdot \dfrac{dP_1}{dy_2} - h_2 \cdot \dfrac{dP_2}{dy_2} = 0.
\end{cases}
$$

In matrix notation, we write:

$$
\begin{bmatrix} \dfrac{dP_1}{dy_1} & \dfrac{dP_2}{dy_1} \\ \dfrac{dP_1}{dy_2} & \dfrac{dP_2}{dy_2} \end{bmatrix} \begin{bmatrix} h_1 \\ h_2 \end{bmatrix} = \begin{bmatrix} \dfrac{dC}{dy_1} \\ \dfrac{dC}{dy_2} \end{bmatrix}
$$

or

$$
\Lambda \times \mathrm{H} = \mathrm{C}.
$$

The solution is $\Lambda^{-1} \times \mathrm{C}$. Certainly, in the Vasicek model, this is also true but because r is in the equation, this step is redundant. Note that we need two bonds with different maturities. It is also seen that the second asset does not have to be a bond, it can be a futures contract, or any other interest rate security. Actually, since the option formula in the CIR model is expressed directly in terms of the state variable, we can pick any two assets. It is just more conventional to use the underlying asset as one hedging asset.

The above example can be extended easily to any number of factors. The matrix notation is quite convenient. The solution would remain the same; only the orders are different. In the following, we give an example of how a hedge is actually and it is also compared to the hedge with the Black model which is a common model in industry.

6.5 TIME-DEPENDENT HEDGING

Time-dependent parameter models need numerical methods to compute prices, and therefore need numerical methods to compute hedges. All numerical methods are based upon the state variable(s) because its distribution is assumed known. As a consequence, it is impossible for an option to have a closed-form hedge ratio in terms of its underlying bond. The hedge needs to be obtained from taking derivatives with respect to the short rate. In a single-factor setting, once a numerical method, say the finite difference method, is set up, all numerical derivatives with respect to the interest rate can be easily computed. The problem is the precision. Especially for options where numbers are small, tiny numbers need double precision to catch which might be memory consuming. And the hedge can be formed by:

$$
\begin{aligned}
V &= C - h \cdot P \\
\frac{dV}{dr} &= \frac{dC}{dr} - h \cdot \frac{dP}{dr} = 0 \\
h &= \frac{dC}{dr} \Big/ \frac{dP}{dr}.
\end{aligned}
$$

Again, the use of the bond price is just by convention and is not really necessary in the numerical methods.

6.6 HIGHER-ORDER HEDGING

Higher-order hedges can be formed in a similar manner. If we want an option position to be both Delta-neutral and Gamma-neutral, then we need to use a bond and another option:

$$V = C - h_\delta \cdot P - h_\gamma A$$
$$\frac{dV}{dr} = \frac{dC}{dr} - h_\delta \cdot \frac{dP}{dr} - h_\gamma \cdot \frac{dA}{dr} = 0$$
$$\frac{d^2V}{dr^2} = \frac{d^2C}{dr^2} - h_\delta \cdot \frac{d^2P}{dr^2} - h_\gamma \cdot \frac{d^2A}{dr^2} = 0.$$

The two hedge ratios can be solved in the same way. If there are two factors, then there would have to be four securities to complete the hedge:

$$V = C - h_1 \cdot A_1 - h_2 A_2 - -h_3 \cdot A_3 - h_4 A_4$$
$$\frac{dV}{dy_1} = \frac{dC}{dy_1} - h_1 \cdot \frac{dA_1}{dy_1} - h_2 \cdot \frac{dA_2}{dy_1} - h_3 \cdot \frac{dA_3}{dy_1} - h_4 \cdot \frac{dA_4}{dy_1} = 0$$
$$\frac{d^2V}{dy_1^2} = \frac{d^2C}{dy_1^2} - h_1 \cdot \frac{d^2A_1}{dy_1^2} - h_2 \cdot \frac{d^2A_2}{dy_1^2} - h_3 \cdot \frac{d^2A_3}{dy_1^2} - h_4 \cdot \frac{d^2A_4}{dy_1^2} = 0$$
$$\frac{dV}{dy_2} = \frac{dC}{dy_2} - h_1 \cdot \frac{dA_1}{dy_2} - h_2 \cdot \frac{dA_2}{dy_2} - h_3 \cdot \frac{dA_3}{dy_2} - h_4 \cdot \frac{dA_4}{dy_2} = 0$$
$$\frac{d^2V}{dy_2^2} = \frac{d^2C}{dy_2^2} - h_1 \cdot \frac{d^2A_1}{dy_2^2} - h_2 \cdot \frac{d^2A_2}{dy_2^2} - h_3 \cdot \frac{d^2A_3}{dy_2^2} - h_4 \cdot \frac{d^2A_4}{dy_2^2} = 0.$$

So, as we can see, the hedge can be rather easy, as long as there are enough traded securities. These high-order hedges can also be used with the time-dependent parameter models. Of course, with closed-form models, derivatives are easier and faster.

6.7 AN EXAMPLE

In a recent paper, Chen and Scott (1995) demonstrate how to conduct hedges with a two-factor CIR model and compare them with those of the Black model. They use the parameter estimates of their Kalman filter paper to build the model. The example they used was hedging Eurodollar futures option with three months to expiration and five years to expiration. The Eurodollar futures option has the underlying asset of LIBOR which deviates from the equivalent domestic treasury rate. Therefore, there is an additional spread variable to capture this extra randomness. Assuming a square root process for the spread they report the parameter estimates as follows:

κ	7.6
θ	0.00125
υ	4%
λ	0

The futures option expires in three months. The futures price from the model is 92.3076 and the price of the call with a strike price of 92.50 is 0.3582. The values for the hedge ratios are $89.604 (face value is $100) long on the three-month T Bill, $46.1696 short on the five-year bond and .04339 short futures contracts. The Black's hedge ratio is 0.5863 which is different from the hedge of the three-factor model. This is of course the three-factor Chen–Scott model uses three different instruments while the Black's model does not differentiate among risks from various sources.

Although Black's model is overly simplistic, the hedging performance of two different models need to be tested before we can draw any favorable conclusion of the multi-factor interest rate model.

CHAPTER 7
CURRENT PROBLEMS AND FUTURE RESEARCH

7.1 INTRODUCTION

Along with the rapid growth in interest rate derivatives, the demand for a "good" term structure model has become the major task of both academics and practitioners. A "good" term structure model needs to be able to accomplish two major missions. First, it needs to be able to fit the current market data. It should not allow arbitrage opportunities. Second, it needs to be able to imply fundamental economic conditions. After all, interest rates are not exogenous financial variables and should be determined endogenous within the economy (Cox, Ingersoll, and Ross (1985a)). In other words, a good term structure model should be able to fit both cross-sectional as well as time-series data.

Unfortunately, after so many models have been developed (reviewed in Chapter 2), none of the them can satisfy both needs. Fixed parameter models provide interpretations of the economy but they cannot fit the traded prices. On the other hand, time-dependent models fit the prices but carry no economic meanings. Furthermore, most time-dependent models are single factor models; therefore they all suffer from a significant drawback — perfect correlation among bond returns of different maturities. This problem can be serious if derivative contracts are written on interest rate differentials. The quality option in T Bond futures, for example, can be viewed as an exchange option which is an option on the difference between two different-maturity bonds. Differential swaps form another example which is a direct option on rate differentials. Certainly, time-dependent models can easily include multiple factors (see Hull and White (1990)) but the models become too complex to manage.

Another serious problem exists in current models is that all current models assume interest rates follow stationary processes while empirically it has been proven that they are actually non-stationary and present cointegration properties. All models need assumptions of the state variables. But if the assumption violates the most important characteristic of the variable, then it can never explain prices correctly. This is like the Black–Scholes formula not being able to explain the *smile* problem and the term structure of the implied volatility because it ignores the volatility which is changing over time. In the term structure, maybe this is the reason why fixed parameter models cannot fit the prices and the time-dependent models need re-parametization constantly. However, so far, the empirical models seen are not easy to be used to develop theoretical models. But that is exactly the reason why this research is so fascinating, because it is still growing.

The effort of building a better term structure model has been under two different directions. One is to keep modifying current method until we reach a satisfactory solution. Chen and Yang (1995) have discovered a clever way to put both fixed parameter model and time-dependent model together. In their model, Chen and Yang propose a flexible fixed parameter term structure model under the general equilibrium environment by CIR. The model has several advantages over previous models. First, it can be made to fit perfectly any curve selected from the markets, while it remains a general equilibrium model that is consistent with the time-series properties of any interest rate. Second, its solution remains in closed-form and so are other European contracts, at least for the Vasicek-based and CIR-based models. Third, it can be viewed as the special case of previous time-dependent parameter models without any loss of their fitting ability. The capability of combining both time-series and cross sectional information is the main contribution of the model.

Another direction is to look at the fundamental characteristics of the interest rates, such as the cointegration characteristic, the conditional heteroskedasticity characteristic, the non-normality characteristic, the long memory characteristic, and the changing risk premia characteristic and develop a better model. I have known some authors working in this direction but have not seen any complete work. The major hurdle of developing a pricing model from this approach is that computation becomes formidable — there is no closed-form solution and sometimes no lattice model. For example, the GARCH option model by Duan (1995) although solves the smile and term structure problem of the implied volatility of the equity options, computing an option price becomes prohibitively difficult and slow.

7.2 THE CURRENT PROBLEMS

The fact that fixed parameter models cannot fit the traded prices is clear. Therefore in this section, we shall concentrate on why we claim that time-dependent models cannot be used as real equilibrium models.

Take Hull–White's extended Vasicek model for example, the short rate can be written, as in Chapter 2, as follows:

$$
\begin{aligned}
dr &= \alpha(t)[\mu(t) - r]dt + \sigma(t)dW \\
&= [\alpha(t)\mu(t) - \alpha(t)r - \sigma(t)\varsigma(t)]dt + \sigma(t)d\hat{W} \\
&= [q(t) - \alpha(t)r]dt + \sigma(t)d\hat{W}.
\end{aligned}
$$

To fit the yield curve and volatility curve, only two parameters are necessary to be time-dependent. From Chapter 2, we know that $\mu(t)$ does not enter into the option formula; therefore it cannot be used to fit the volatility curve. On the other hand, either $\alpha(t)$ or $\sigma(t)$ can be used to fit the volatility curve since both enter into the option formula. Since the drift parameter does not appear in the option formula, the fitting of the yield curve can be made after the fitting of the volatility. In other words, we can determine $\alpha(t)$ or $\sigma(t)$ first using the option prices and then plug it into the bond pricing function to determine $\mu(t)$ by making the model price the same as the market price. Hull and White (1990, 1993) have shown that for a lattice to recombine, σ in the state variable would need to be constant. Therefore, it is natural to choose α to be the parameter for the fitting of the volatility.

To see that, let us assume a three-period model as follows:

The bond expires at the end of the third year. We assume that options are traded on all possible bond contracts. That is why we need option prices of $C(0,1,3)$ and $C(0,2,3)$. Within every period, the parameter value is constant. Therefore, we can use the simple Vasicek/Jamshidian model to find the parameter value. For example, suppose we observe the following prices:

current time	option maturity	bond maturity	option prices
0	1	3	.00364
0	2	3	.08500

current time	bond maturity	bond prices
0	1	.95
0	2	.90
0	3	.85

σ is set to be 3% and the strikes are 0.9 and 0.85 respectively.

For each option price, recall the pricing formula is:

$$C(t,T,s) = P(t,s)N(d) - P(t,T)KN(d - \sqrt{V_p})$$

where

$$V_p = \sigma^2 \phi(T)^2 \left[\int_t^T \left(\phi^{-1}(w)\right)^2 dw \right] \left[\int_T^s \phi(u) du \right]^2.$$

Therefore, we need to solve the simultaneous equations for the two α's. The way to do that is to first convert the integrals into Reiman sums and then solve for α's. The result are 1 for the first year and 0.8 for the second year respectively. Plug this set of results into the bond equation and match for real bond prices, we can then solve for the two μ's to be 6% and 6.2% respectively.

We can see that when the volatility curve is long, the solving of the time-dependent parameter α becomes rather difficult because we need to solve for the n-dimensional simultaneous equations. In all published papers, we normally see volatility curves given as a black box. No one ever questions how these curves are obtained. Note that the state variable is not directly observed, nor is its volatility over time.

Using traded options with various times to maturity (on the same bond) is certainly one way of obtaining the volatility curve and the time-dependent parameter. However, it should be noted that if the model is incorrect, then when we use another series of option prices (another option on a different maturity bond), we shall arrive at a different set of the time-dependent parameter values. This potentially inconsistency within the model opens a way to test if a term structure model is appropriately specified.

Due to the difficult solving process of the volatility curve, Chen and Yang (1995) recently proposed a model that decomposes the interest rate dynamics into different-purpose state variables. This approach has several advantages over the previous models. We shall demonstrate their model in the next section.

7.3 THE CHEN–YANG MODEL

They decompose the instantaneous short rate into three independent variables:

$$r(t) = y_1(t) + y_2(t) + y_3(t)$$

where

$$\begin{cases} dy_1 = d\mu_1(t) \\ dy_2 = \sigma_2(t)d\hat{W}_2 \\ dy_3 = [\alpha(\mu_3 - y_3) - \sigma_3\varsigma]dt + \sigma_3 d\hat{W}_3. \end{cases}$$

The pure discount bond price is therefore:

$$\begin{aligned} P(t,T) &= \hat{E}_t\left[\exp\left(-\int_t^T r(s)du\right)\right] \\ &= \exp\left(-\int_t^T y_1(s)du\right)\hat{E}_t\left[\exp\left(-\int_t^T y_2(s)du\right)\right]\hat{E}_t\left[\exp\left(-\int_t^T y_3(s)du\right)\right] \\ &= P_1(t,T)P_2(t,T)P_3(t,T). \end{aligned}$$

While P_3 is the standard Vasicek model, $P_2(t,T)$ can be solved in the following:

$$\begin{aligned} P_2(t,T) &= \hat{E}_t\left[\exp\left(-\int_t^T y_2(s)ds\right)\right] \\ &= e^{-m+V/2} \end{aligned}$$

where

$$m(t,T) = y_2(t)(T-t) \text{ and}$$
$$V(t,T) = 2\int_t^T \int_t^s \int_t^u \sigma_2(w)^2 dwduds.$$

The option formula is:

$$C(t,T,\tau) = P(t,\tau)N(d) - P(t,T)KN(d - \sqrt{V_p})$$

where

$$d = \frac{\ln(P(t,\tau)/KP(t,T))}{\sqrt{V_p}} + \frac{V_p}{2}$$

and

$$\begin{aligned} V_p &= \text{var}(\ln P_2(T,\tau) + \ln P_3(T,\tau)) \\ &= \text{var}(\ln P_2(T,\tau)) + \text{var}(\ln P_3(T,\tau)). \end{aligned}$$

Each variance component is computed by the standard method using the results given in Chapter 2. For the variance of bond 3:

$$\begin{aligned} \text{var}(\ln P_3(T,\tau)) &= \text{var}(-r(T)F(T,\tau)) \\ &= F(T,\tau)^2\, \text{var}(r(T)) \\ &= F(T,\tau)^2 \sigma_3^2\left(1 - e^{-2\alpha(T-t)}\right)/2\alpha. \end{aligned}$$

The other variance, the variance of $\ln P_2$ can be easily solved by:

$$\begin{aligned}\text{var}(\ln P_2(T,\tau)) &= \text{var}(y_2(T)(\tau - T)) \\ &= (\tau - T)^2 \int_t^T \sigma_2(s)^2 ds \\ &= (\tau - T)^2 I(t,T).\end{aligned}$$

Chen and Yang point out that it is exactly the $I(\cdot,\cdot)$ function that makes their model differ from the other models computationally. Because this function is approximately a linear function in time, when time intervals are small, the volatility curve can be cut into several pieces each of which is simply an $I(\cdot,\cdot)$. There is no need to solve any simultaneous equations and the volatility curve can directly be taken from traded option prices.

Although the Chen and Yang model is computationally easier, it violates the variance condition when one wants to build a lattice model — the volatility parameter cannot be time-dependent, argued by Hull and White (1990b, 1993). However, Chen and Yang argue that this is not a problem because the second variable has no mean reversion. They argue that non-constant volatility parameter would pose a problem only when one wants both the mean reversion and volatility to be time-dependent. Then the lattice would face non-recombination problems.

They then go on to test their model against the Hull–White model in an attempt to show that the perfect correlation problem in all single factor models would produce significantly biased results. And the ability of their model to utilize the closed-form solution can further produce more accurate prices. Although their model needs to be further tested for its realiability, the computational efficiency and the closed-form feature of the model can be attractive to practitioners and academics.

7.5 FUTURE RESEARCH

As discussed in the Introduction of this chapter, the current models are far from satisfaction. Chen and Yang's attempt is certainly admirable and does solve the problem of fitting and provide computational efficiency. But whether it is closer to *the true* model still needs to be tested. Specifically, the volatility curve comes from one particular bond! It is known that the volatility curve should come from the state variable. Since the state variable is neither traded nor observable, certain bond's options are used to retrieve the volatility curve for the state variable. However, we have multiple bonds traded in the marketplace everyday, why pick a specific bond and not others for the volatility curve? If the model is truly correct, then the volatility curve coming from one bond using the model should be consistent with the volatility curve coming from another bond using the same model. This certainly opens up a way to test if the model is correctly specified. A similar example is the "smile" problem in the Black and Scholes model. The Black–Scholes model generates different implied standard deviations for different strikes. This is a direct rejection of the model because the underlying asset is the same and the time to maturity of these options is the same; so there is no reason why the same stock over the same investment horizon should have more than one volatility estimate. In the term structure theory, we would hope to obtain a model that can truly reflect all important characteristics of our economy.

Empirical evidence recently shows that interest rates are cointegrated and present heteroskdesticity . Some researchers even believe that interest rates have long memory. If so, then it is very likely that none of the models is suitable for pricing interest rate derivatives. For example, Chan *et al.*, when fitting the short rate data through a constant elasticity variance (or CEV) model find that there is no structure break in October of 1979. Chung and Hung (1995) argue that this is because the heteroskdesticity structure in the CEV model is too simplistic. It is important to know that parameters that treat the structure break as non-existent will be different from those that recognize the structure break. If the problem is truly caused by the structure of heteroskdesticity, then the CIR model is not suitable. And do not forget that Gaussian-based models assumes no heteroskdesticity.

Cointegration is another problem. It is known that if we run a regression of one random walk variable on another random walk variable, we get all kinds of strange results. If we note the regression results given in Chapter 2 by Chen and Yang, the short rate cannot pass the unit root test. This immediately imply that the speed parameter, α, is very unstable. It can change from one number to a very different number if we change the estimation period. And we know that it is important in pricing options because it enters the volatility calculation.

Long memory poses another big issue for us. It will destroy the complete market assumption (in continuous time) and we will not obtain any easy solution, even for the simplest European style. Since the evidence is limited, its impact to option pricing is not yet clear. Nonetheless, current models cannot capture this feature.

APPENDIX

A. Perfect Correlation

The problem of all single factor models is that all bond yields are perfectly correlated. Define the yield to maturity as:

$$y(t,T) = -\frac{\ln P(t,T)}{T-t}.$$

Therefore,

$$\begin{aligned}
\rho_t(y(s,T_1),y(s,T_2)) &= \frac{\operatorname{cov}(y(s,T_1),y(s,T_2))}{\sqrt{\operatorname{var}(y(s,T_1))\operatorname{var}(y(s,T_2))}} \\
&= \frac{\operatorname{cov}\left(\frac{-\ln P(s,T_1)}{T_1-s},\frac{-\ln P(s,T_2)}{T_2-s}\right)}{\sqrt{\operatorname{var}\left(\frac{-\ln P(s,T_1)}{T_1-s}\right)\operatorname{var}\left(\frac{-\ln P(s,T_2)}{T_2-s}\right)}} \\
&= \frac{\operatorname{cov}(r(s),r(s))}{\operatorname{var}(r(s))} \\
&= 1.
\end{aligned}$$

BIBLIOGRAPHY

Abken, P., 1989, "Valuing Default-Risky Interest Rate Caps: A Monte Carlo Approach," Working Paper, Federal Reserve Bank of Atlanta, November.

Arak, M. and L. Goodman, 1987, "Treasury Bond Futures: Valuing the Delivery Option," *Journal of Futures Markets.*

Arnold, Ludwig, 1974, *Stochastic Differential Equations: Theory and Applications*, John Wiley and Sons.

Bachelier, L., 1964, *Théorie de la Speculation*, reprinted in English in the Radom Characters of Stock Market Prices, Cambridge, MIT.

Ball, C., 1989, "Estimation of a Diffusion Process for Spot Rates," Working Paper, University of Michigan, August.

Ball, C. and W. Torous, 1983, "Bond Price Dynamics and Options," *Journal of Financial and Quantitative Analysis*, December.

Ball, C. and W. Torous, 1986, "Futures Options and the Volatility of Futures Prices," *Journal of Finance*, September.

Barone-Adesi, G. and R. Whaley, 1987, "Efficient Analytic Approximation of American Option Values," *Journal of Finance*, Vol. 42, No. 2, p. 301–320.

Beaglehole, D. and M. Tenny, 1992, "Corrections and Additions to 'A Nonlinear Equilibrium Model of the Term Structure of Interest Rates,'" *Journal of Financial Economics*, Vol. 32, p. 345–353.

Beckers, S., 1980 (June), "The CEV Model and Its Implications for Option Pricing," *Journal of Finance*, Vol. 35, p. 661–673.

Benninga, S. adn M. Smirlock, 1985, "An Empirical Analysis of the Delivery Option, Marking to Market, and the Pricing of Treasury Bond Futures," *Journal of Futures Markets.*

Bicksler, J. and A. Chen, 1986 (July), "An Economic Analysis of Interest Rate Swaps," *Journal of Finance*, Vol. 41, No. 3, p. 645–655.

Black, F., 1976, "The Pricing of Commodity Contracts," *Journal of Financial Economics*, p. 167–179.

Black, F., E. Derman, and W. Toy, 1990 (January–February), "A One-Factor Model of Interest Rates and Its Application to Treasury Bond Options," *Financial Analyst Journal*, p. 33–39.

Black, F. and P. Karasinski, 1991 (July–August), "Bond and Option Pricing When Short Rates Are Lognormal," *Financial Analysts Journal*, Vol. 47, p. 52–59.

Black, F. and M. Scholes, 1973, "The Pricing of Options and Corporate Liabilities,"

Journal of Political Economy, p. 637–654.

Boyle, P., 1980, "Recent Models of the Term Structure of Interest Rates with Actuarial Applications," *Transactions of the International Congress of Actuaries.*

Boyle, P., 1989 (March), "The Quality Option and Timing Option in Futures Contracts," *Journal of Finance*, Vol. 44, No. 1.

Brennan, M. and E. Schwartz, 1978a, "A Continuous Time Approach to the Pricing of Bonds," *Journal of Banking And Finance*, p. 133–155.

Brennan, M. and E. Schwartz, 1978b (September), "Finite Difference Methods And Jump Processes Arising in the Pricing of Contingent Claims: A Synthesis," *Journal of Financial And Quantitative Analysis.*

Brennan, M. and E. Schwartz, 1982 (September), "An Equilibrium Model of Bond Pricing And A Test of Market Efficiency," *Journal of Financial And Quantitative Analysis.*

Broadie, M. and S. Sundaresan, 1987, "The Pricing of Timing and Quality Options: An Application to Treasury Bond Futures Markets," Working Paper.

Brown, S. and P. Dybvig, 1986 (July), "The Empirical Implications of the Cox, Ingersoll, and Ross Theory of the Term Structure of Interest Rates," *Journal of Finance.*

Campbell, J., 1986 (March), "A Defense of traditional Hypotheses about the Term Structure of Interest Rates," *Journal of Finance.*

Capozza, D. R. and B. Cornell, 1979, "Treasury Bill Pricing in the Spot and Futures Markets," *The Review of Economics and Statistics*, p. 513–520.

Carr, P., 1988 (February), "Valuing Bond Futures and The Quality Option," Working Paper, University of California at Los Angelos.

Chance, D. and M. Hemler, 1993, "The Impact of Delivery Options on Futures Prices: A Survey," *Journal of Futures Markets*, Vol. 13, No. 2, p. 127–156

Chaplin, G., 1987a, "A Formula for Bond Option Values Under An Ornstein–Uhlenbeck Model for the Spot," Working Paper No. 87–15, University of Waterloo.

Chaplin, G., 1987b, "The Term Structure of Interest Rates: A Model Based on Two Correlated Stochastic Processes with Closed Form Solutions for Bond and Option Prices," Working Paper No. 87–16, University of Waterloo.

Chen, R., 1991 (September), "Pricing Stock and Bond Options when the Default-Free Rate is Stochastic: A Comment," *Journal of Financial and Quantitative Analysis.*

Chen, R., 1992 (March), "Exact Solutions for Futures and European Futures Options on Pure Discount Bonds," *Journal of Financial and Quantitative Analysis*, Vol. 27, No. 1, p. 97–107.

Chen, R., 1995 (April), "Two Factor, Preference Free Formulas for Interest Rate Sensitive Claims," *Journal of Futures Markets*.

Chen, R. and L. Scott, 1992, "Pricing Interest Rate Options in a Two-Factor Cox–Ingersoll–Ross Model of the Term Structure," *Review of Financial Studies*, Vol. 5, No. 4, p. 613–636.

Chen, R. and L. Scott, 1993, "Maximum Likelihood Estimation of a Multi-Factor Equilibrium Model of the Term Structure of Interest Rates," *Journal of Fixed Income*, Vol. 3, No. 3, December, p. 14–32.

Chen, R. and L. Scott, 1995 (Winter), "Interest Rate Options in Multi-Factor CIR Models of the Term Structure," *Journal of Derivatives*.

Chen, R. and L. Scott, 1995b, "Multi-Factor Cox–Ingersoll–Ross Models of the Term Structure of Interest Rates: Estimates and Tests Using a Kalman Filter," Working Paper, Rutgers University and University of Georgia.

Chen, R. and T. Yang, 1995, "The Relevance of Interest Rate Processes in Pricing Mortgage-Backed Securities," *Journal of Housing Research*, Vol. 6, No. 2.

Chung, C. and M. Hung, 1995, "A General Model for Short Term Interest Rates," Working paper, Michigan State University.

Clewlow, L. and C. Strickland, 1994, "A Note on Parameter Estimation in the Two-Factor Longstaff and Schwartz Interest Rate Model," *Journal of Fixed Income*, Vol. 3, No. 4, March, p. 95–100.

Constantinides, G., 1992, "A Theory of the Nominal Term Structure of Interest Rates," *Review of Financial Studies*, Vol. 5, No. 4, p. 531–552.

Cooper I. and A. Mello, "1991 (June), "The Default Risk of Swaps," *Journal of Finance*, Vol. 48, No. 2, p. 597–620.

Cornell, B and M Reinganum, 1981 (December), "Forward and Futures Prices: Evidence from the Foreign Exchange Markets," *Journal of Finance*.

Courtadon, G., 1982, "The Pricing of Options on Default-free Bonds," *Journal of Financial and Quantitative Analysis*, Vol. 17, p. 75–100.

Cox, J., 1975 (November), "Notes on Option Pricing I: Constant Elasticity of Diffusions," Unpublished draft, Stanford University.

Cox, J., J. Ingersoll, and S. Ross, 1980 (May), "An Analysis of Variable Loan Contracts," *Journal of Finance*.

Cox, J., J. Ingersoll, and S. Ross, 1981, "The Relation Between Forward Prices and Futures Prices," *Journal of Financial Economics*, p. 321–346.

Cox, J., J. Ingersoll, and S. Ross, 1985a (March), "An Intertemporal General Equilibrium Model of Asset prices," *Econometrica*, Vol. 53, No. 2, p. 363–384.

Cox, J., J. Ingersoll, and S. Ross, 1985b (March), "A Theory of The Term Structure of Interest Rates," *Econometrica*, Vol. 53, No. 2, p. 385–407.

Cox, J. and S. Ross, 1976, "The Valuation of Options for Alternative Stochastic Processes," *Journal of Financial Economics*, p. 145–166.

Cox, J, S. Ross, and M. Rubinstein, 1979, "Option Pricing: A Simplified Approach," *Journal of Financial Economics*, Vol. 7, p. 229–63.

Dothan, L., 1978, "On the Term Structure of Interest Rates," *Journal of Financial Economics*, p. 59–69.

Dybvig, P., 1989 (February), "Bond and Bond Option Pricing Based on the Current Term Structure," Working Paper, Washington University.

Dybvig, P., J. Ingersoll, and S. Ross, 1987, "Long Rates Can Never Fall," working paper, Yale University.

Emanuel, D. and J. MacBeth, 1982 (November), "Further Results on the Constant Elasticity of Variance Call Option Pricing Formula," *Journal of Financial and Quantitative Analysis*, Vol. 17, p. 533–554.

Engle, R. and M. Watson, 1981, "A One Factor Multivariate Time Series Model of Metropolitan Wage Rates," *Journal of the American Statistical Association*, Vol. 76, p. 774–781.

Fama, E., 1965 (January), "The Behavior of Stock Market Prices," *Journal of Business*.

Fama, E., 1970 (May), "Efficient Capital Markets: A Review of Theory and Empirical Work," *Journal of Finance*, Vol. 25, p. 383–417.

Fama, E., 1976, *Foundation of Finance*, Basic Books Inc. Publishers.

Feldman, D., 1992, "European Options on Bond Futures: A Closed Form Solution," *Journal of Futures Markets*.

Feller, W., 1951 (July), "Two Singular Diffusion Problems," *Annals of Mathematics*, Vol. 54, No. 1.

Feller, W., 1968 , *An Introduction to Probability Theory and Its Applications*, Vol. 1.

Feller, W., 1971, *An Introduction to Probability Theory and Its Applications*, Vol. 2.

Fisher, S., 1978 (March), "Call Option Pricing When the Exercise Price is Uncertain, and the Valuation of Index Bonds," *Journal of Finance*, Vol. 33, No. 1, p. 169–176.

Flesaker, B., 1989 (March), "Arbitrage Free Pricing of Interest Rate Futures And Forward Contracts," Working Paper, UC Berkeley.

French, K., 1983, "A Comparison of Futures and Forward Prices," p. 311–342, *Journal of Financial Ecomomics*.

Friedman, Arvner, 1975, *Stochastic Differential Equations and Applications*, Academic Press.

Garman M. and S. Kolhegen, 1983 (December), "Foreign Currency Option Values," *Journal of International Money and Finance*, p. 231–7.

Gay, G. and S. Manaster, 1986 (May), "Implicit Delivery Options and Optimal Delivery Strategies for Financial Futures Contracts," *Journal of Financial Economics*, Vol. 16.

Geske, R., (1979) "The Valuation of Compound Options," *Journal of Financial Economics*, p. 63–81.

Geske, R., 1977, "The Valuation of Corporate Liabilities as Compound Options," *Journal of Financial and Quantitative Analysis*, Vol. 12, p. 541–552.

Gibbons, M. and K. Ramaswamy, 1993, "A Test of the Cox, Ingersoll, and Ross Model of the Term Structure," *Review of Financial Studies*.

Harrison J. and D. Kreps, 1979, "Martingales and Arbitrage in Multi-Period Securities Markets," *Journal of Economic Theory*, p. 381–408.

Harrison J. and S. Pliska, 1981, "Martingales and Stochastic Integrals in the Theory of Continuous Trading," *Stochastic Processes and Their Applications*, p. 215–260.

Haugen, R., 1986, *Modern Investment Theory*, Prentice Hall.

Heath, D., R. Jarrow, and A. Morton, 1992 (February) "Bond Pricing and the Term Structure of Interest Rates: A New Methodology," *Econometrica*.

Hedge, S., 1988, "An Empirical Analysis of Implicit Delivery Options in the Treasury Bond Futures Contract," *Journal of Banking and Finance*, Vol. 12, p. 469–492.

Hemler, M., 1990 (December), "The Quality Delivery Option in Treasury Bond Futures Contracts," *Journal of Finance*, Vol. 45, No. 5.

Heston, S., 1989, "Testing Continuous Time Models of the Term Structure of Interest Rates: A New Methodology for Contingent Claims Valuation," Working Paper, Carnegie Mellon University.

Heston, S., 1993, "A Closed Form Solution for Options with Stochastic Volatility with Applications to Bond and Currency Options," *Review of Financial Studies*, Vol. 6, No. 2.

Ho, T. and S. Lee, 1986 (December), "Term Structure Movements and Pricing Interest Rate Contingent Claims," *Journal of Finance*.

Hsin, C., 1990, "Investigations on Options on Default-Free Bond Futures," Unpublished Ph.D. Dissertation, University of Illinois.

Hull, J. and A. White, 1987 (March), "The Pricing of Options on Assets with Stochastic Volatilities," *Journal of Finance*, p. 281–300.

Hull, J. and A. White, 1990a, "Pricing Interest Rate Derivative Securities," *Review of Financial Studies*, Vol. 3, No. 4, p. 573–592.

Hull, J. and A. White, 1990b, "Valuing Derivative Securities Using the Explicit Finite Difference Method," *Journal of Financial and Quantitative Analysis*, Vol. 25, No. 1, p. 87–100.

Hull, J. and A. White, 1993, "One Factor Interest Rate Models and the Valuation of Interest Rate Derivative Securities," *Journal of Financial and Quantitative Analysis*, Vol. 28,

Hull, J., 1993, *Options, Futures, and Derivative Securities*, Prentice Hall, 2nd ed.

Ingersoll, J., 1987, *Theory of Financial Decision Making*, Rowman and Littlefield.

Jacklin, C. and M. Gibbons, 1989 (December), "CEV Diffusion Estimation," Working Paper, Stanford University.

Jamshidian, F., 1987, "Pricing of Contingent Claims in The One-Factor Term Structure Model," Working Paper, Merrill Lynch Capital Markets.

Jamshidian, F., 1989a (March), "An Exact Bond Option Formula," *Journal of Finance*.

Jamshidian, F., 1989b (March), "Closed Form Solution for American Options on Coupon Bonds in the General Gaussian Interest Rate Model," Working Paper, Merrill Lynch Capital Markets.

Jamshidian, F., 1989b (October), "The Multifactor Gaussian Interest Rate Model and Implementation," Working Paper, Merrill Lynch Capital Markets.

Jamshidian, F., 1993, "Price Differentials," RISK, 6:7, p. 48–51.

Johnson, N. and S. Kotz, 1970, *Distributions in Statistics: Continuous Univariate Distributions — 2*, Boston: Houghton Mifflin Company.

Jones, R., 1986, "Conversion Factor Risk in Treasury Bond Futures: Comment," *Journal of Futures Markets*, Vol. 5.

Kane, A. and A. Marcus, 1986 (Summer), "The Quality Option in the Treasury Bond Futures Contract: An Empirical Assessment," *Journal of Futures Markets*, Vol. 6.

Kane, A. and A. Marcus, 1986, "Valuation and Optimal Exercise of the Wild Card Option in the Treasury Bond Futures Contracts," *Journal of Futures Markets*, Vol. 41.

Karlin, S. and H. Taylor, 1981, *A Second Course in Stochastic Processes*, Academic Press.

Kilcollin, T., 1982, "Difference Systems in Financial Futures Markets," *Journal of Finance*, Vol. 37.

Kolb, R., 1994, *Futures, Options, and Swaps*, Kolb Publishing Company.

Kraus, A., and M. Smith, 1993 (June), "A Simple Mulit-factor Term Structure Model," *Journal of Fixed Income*, p. 19–23.

Langetieg, T., 1980, "A Multivariate Model of the Term Structure," *Journal of Finance*,

Vol. 15, No. 1, p. 71–97.

Lauterbach, B. and P. Schultz, 1990 (September), "Pricing Warrants: An Empirical Study of the Black–Scholes Model and Its Alternatives," *Journal of Finance*, Vol. 45, p. 1081–1120.

Litterman, R. and J. Scheinkman, 1991, "Common Factors Affecting Bond Returns," *Journal of Fixed Income Securities*, Vol. 1, No. 1, p. 54–61.

Litzenberger, R., 1992 (July), "Swaps: Plain and Fanciful," *Journal of Finance*, Vol.47, No. 3, p. 831–850.

Livingston, Miles, 1987 (March), "The Delivery Option on Forward Contracts," *Journal of Financial and Quantitative Analysis*, Vol. 22.

Longstaff, F., 1989, "A Non-linear General Equilibrium Model of the Term Structure of Interest Rates," *Journal of Financial Economics*, Vol. 23, p. 195–224.

Longstaff, F. and E. Schwartz, 1992 (December), "Interest Rate Volatility and the Term Structure: A Two Factor General Equilibrium Model," *Journal of Finance*, Vol. 47, No. 4, p. 1259–82.

Longstaff, F. and E. Schwartz, 1993 (September), "Implementation of the Longstaff–Schwartz Interest Rate Model," *Journal of Fixed Income*, p. 7–14.

MacBeth, J. and L. Merville, 1979 (December), "An Empirical Examination of the Black–Scholes Call Option Pricing Model," *Journal of Finance*, Vol. 34, p. 1173–1186.

MacBeth, J. and L. Merville, 1980 (May), "Tests of the Black–Scholes and Cox Call Option Valuation Models," *Journal of Finance*, Vol. 35, p. 285–301.

Margrabe, W., 1978, "The Value of an Option to Exchange One Asset for Another," *Journal of Finance*, Vol. 33, p. 177–186.

Merton, R., 1971, "Optimal Consumption and Portfolio Rules in a Continuous Time Model," *Journal of Economic Theory*, p. 373–413.

Merton, R., 1973a, "An Intertemporal Asset Pricing Model," *Econometrica*, p. 867–887.

Merton, R., 1973b, "The Rational Theory of Option Pricing," *Bell Journal of Economics and Management Science*, Vol. 4, p. 141–183.

Merton, R., 1976, "Option Pricing When Underlying Stock Returns Are Discontinuous," *Journal of Financial Economics*, p. 125–144.

Pearson, N. and T. Sun, 1994, "A Test of the Cox, Ingersoll, and Ross Model of the Term Structure of Interest Rates Using the Method of Maximum Likelihood," *Journal of Finance*.

Rabinovitch, R., 1989 (December), "Pricing Stock and Bond Options when the Default-Free Rate is Stochastic," *Journal of Financial and Quantitative Analysis*.

Richard, S., 1978, "An Arbitrage Model of the Term Structure of Interest Rates," *Journal of Financial Economics*, p. 33–57.

Ritchken, P. and L. Sankarasubramanian, 1992 (July), "Pricing the Quality Option in T Bond Futures," *Mathematical Finance*, Vol. 2, No. 3, p. 197–214.

Ross, S., "Comment on R. C. Merton's Lecture," *The Geneva Papers on Risk and Insurance*, Vol. 14, p. 271–273, 1989.

Sankaran, M., 1963 (June), "Approximations to the Non-Central Chi-Square Distribution," *Biometrica*, Vol. 50, p. 199–204.

Schaefer S. and E. Schwartz, 1984 (December), "A Two Factor Model of The Term Structure: An Approximate Analytic Solution," *Journal of Financial and Quantitative Analysis*.

Schroder, M., 1989 (March), "Computing the CEV Option Pricing Formula," *Journal of Finance*, Vol. 44, p. 211–220.

Schwartz, E. and W. Torous, 1988, "Prepayment and the Valuation of Mortgage-Backed Securities," Working Paper #11–88, UCLA.

Scott, L., 1987 (December), "Option Pricing When the Variance Changes Randomly: Theory, Estimation, and An Application," *Journal of Financial and Quantitative Analysis*.

Scott, L., 1989 (Spring), "Stock Price Changes with Random Volatility and Jumps: Some Empirical Evidence," *Quarterly Journal of Economics and Business*.

Scott, L., 1993, "Pricing Stock Options in a Jump Diffusion Model with Stochastic Volatility and Interest Rates: Applications of Fourier Inversion Methods," University of Georgia.

Sharp, K., 1987, "Applications of Stochastic Models of Interest Rates," Unpublished Dissertation, University of Waterloo.

Smith, C., 1976, "Option Pricing: A Review," *Journal of Financial Economics*, p. 1–51.

Smith, C., C. Smithson, and L. Wakeman, 1987, "Credit Risk and the Scope of the Regulation of Swaps," Federal Reserve Bank of Chicago.

Stoll, H. and R. Whaley, 1993, *Futures and Option: Theory and Applications*, South-Western.

Stulz, R., 1982, "Options on the Minimum or the Maximum of Two Risky Assets: Analysis and Applications," *Journal of Financial Economics*, Vol. 10, p. 161–185.

Sundaresan, S., 1991, "Valuation of Swaps," *Recent Development of International Banking and Finance*, Chapter 12, p. 407–440.

Titman, S., 1992 (September), "Interest Rate Swaps and Corporate Financing Choices,"

Journal of Finance, Vol. 47, No. 4, p. 1503–1516.

Turnbull S. and F. Milne, 1991"A Simple Approach to Interest Rate Option Pricing," *Review of Financial Studies*, Vol. 4, No. 1, p. 87–120.

Turnbull, S., 1987, "Swaps: A Zero Sum Game?" *Financial Management*, p. 15–21.

Turnbull, S., 1993, "Pricing and Hedging Diff Swaps," Working Paper, Queens University.

Vasicek, O., 1977, "An Equilibrium Characterization of The Term Structure," *Journal of Financial Economics*, p. 177–188.

Wall, L. and J. Pringle, 1988, "Interest Rate Swaps: A Review of Issues," *Economic Review*, Federal Reserve Bank of Atlanta, p. 22–40.

Whittaker, G., 1987, "Pricing Interest Rate Swaps in An Options Pricing Framework," unpublished manuscript, Federal Reserve Bank of Kansas City.

INDEX

T

V

W

Y

p38 HL Any miss in the fitting of the yield curve can cause significant consequences in pricing derivatives.
Since HL model takes bond prices as given, their model cannot be used to price bonds. It is used for IR derivatives s.a. options.

p144 the problem of all single factor models is that all bond yields are perfectly correlated.

p38 HoLee: Time dependent parameter in
$dr = \theta(t)dt + \sigma dW$. No pricing error in fitting to y-curve
Has a closed form solution. θ is cooked so that the discount bond pricing formula is flexible enough to fit every point.
The short rate is normally distributed.

p47 BDT differs from HL in that BDT fit the volatility curve as well. Under BDT the short rate is lognormal.
BK is the continuous-time version of BDT. It is like HW. However BK is lognormal while extended Vasicek is normal.

p53 All time-dependent parameter ~~models~~ models (HL, HW, BDT, BK, ext. Vasicek, HJM) adjust parameters to fit yield & volatility curves.

P13 The discrete forward rate $f(t, T_f, T)$, prevailing at time t, for the period (T_f, T) of length $(T-T_f)$ years, assuming annual compounding is obtained from the price ~~at time~~ $\psi(t, T_f, T)$ prevailing at time t, of the bond from T_f to T by solving

$$\frac{1}{\psi(t,T_f,T)} = [1+ f(t,T_f,T)]^{T-T_f} \quad \text{i.e.}$$

$$1+ f(t,T_f,T) = \sqrt[T-T_f]{\frac{1}{\psi(t,T_f,T)}} \quad \therefore f(t,T_f,T) = \sqrt[(T-T_f)]{\frac{1}{\psi}} - 1$$

Since $\psi(t,T_f,T) = \frac{P(t,T)}{P(t,T_f)}$, $= (1+f)^{T-T_f}$

$$\therefore -(T-T_f)\ln(1+f) = \ln P(t,T) - \ln P(t,T_f)$$

$$\therefore -\ln(1+f) = \frac{\ln P(t,T) - \ln P(t,T_f)}{T-T_f}$$

p16 Duration $D = \frac{-dP}{dy} \cdot \frac{1}{P}$, Mod Dur. $= \frac{-dP}{dy}\frac{(1+y/m)}{P}$

Convexity $= \frac{d^2P}{dy^2}/P$.

p23 For a risk-neutral investor the expected value of stock price equals SR where S is current stock price and R = 1+ riskfree rate.

$\therefore$ In a binary model of stock prices for a risk neutral investor the probability measure (p, 1-p) where p is prob. of upmove to Su & 1-p is prob of downmove to Sd, is obtained by solving $pSu + (1-p)Sd = RS$.

p29 If int. rates can be -ve, bond prices can exceed 1.

p34 HoLee model uncertainty by putting perturbation on forward prices BDT model is richer than HL in that it also fits the volatility curve